Climate Change — Who's Carrying the Burden?

The chilly climates of the global environmental dilemma

Edited by L. Anders Sandberg
and Tor Sandberg

Third volume in the
Our Schools/Our Selves book series
2010

NOV 18 2010

Climate Change — Who's Carrying the Burden?
The chilly climates of the global environmental dilemma
2010

Third volume in the *Our School / Our Selves* book series.

EXECUTIVE EDITOR
Erika Shaker

VOLUME EDITORS
L. Anders Sandberg and Tor Sandberg

EDITORIAL OFFICE
Canadian Centre for Policy Alternatives
Suite 205, 75 Albert Street, Ottawa, ON, K1P 5E7
Tel: (613) 563-1341 Fax: (613) 233-1458

ISBN 978-1-926888-06-4

PRODUCTION
Typesetting and design: Nancy Reid
Cover illustration: Gallery Galschiøt, Odense, Denmark (www.aidoh.dk)
Cover design: Nancy Reid (www.nrgrafix.com)

Printed in Canada by DLR International Printing

Contents

Preface

This book is dedicated to those that suffer the most from climate change yet are the least responsible for it. We hope the message and insights we provide may make a contribution, however small, to the recognition of their situation and to give a voice to their concerns.

Climate change is typically about the devastating impact of global climate change in the form of rising temperatures, more extreme weather, melting polar ice caps, drowning polar bears, rising sea levels, floods and droughts. Climate change is also seen to affect everybody and, therefore, calls for global solutions. The international Kyoto Agreement on carbon emissions constitutes one such global measure. The trade in carbon-emissions permits constitutes another measure; so does the investment in new technologies, ranging from giant mirrors in space that deflect the sun's rays to burying carbon emissions in the ground. At the same time, we are all asked to take individual responsibility for climate change by practicing the three Rs: reuse, reduce, and recycle.

Climate Change — Who's Carrying the Burden? questions these developments by focusing on the distributional impact and visions of climate change and the connection of climate change to wider systemic forces. We join a group of climate justice advocates who are critical of markets, new technologies, and international agreements as solutions to the climate change dilemma. We share their call for exploring the origins of climate change and the places where its impacts are felt the most, such as the Tar Sands of Alberta, the Gulf of Guinea in West Africa, the Canadian North, the coastal regions of Bangladesh, and the island states of the Pacific.

We also propose that the concept of climate change itself can be an oppressive force in not only hiding the historical connections of the carbon economy to colonialism, capitalism, and a rampant and exploitative resource extraction, but also the resiliencies, possibilities and alternatives articulated by groups who fight and stand outside the carbon economy. There are, we argue, chilly climates that surround the discussions on climate change that erase, exclude and marginalize alternative views and possibilities.

The artwork of sculptor and activist Jens Galschiøt that graces the cover and the first page of each chapter of this book illustrates the connection between climate and system change. His sculpture "Justitia, Western Goddess of Justice" was a prominent symbol in Copenhagen during the COP15 Climate Change Conference in December 2009. It was accompanied by the following inscription: "I'm sitting on the back of a man — he is sinking under the burden — I will do everything to help him — except to step down from his back." The inscription refers to the heavy climate change burden carried by the Global South, and the climate debt owed it by a recalcitrant and conspicuously consuming Global North.

What is interesting about the Justitia sculpture, however, is that it was not made for the Copenhagen Climate Change Conference but for the Social Forum in London in 2004 (an offshoot of the World Social Forum in Porto Alegre). At the Forum, Justitia represented the harsh impact imposed by First World agricultural policies on Third World farmers. Galschiøt understands and calls attention to climate change as one phenomenon among many that is contributing to a more unequal and more environmentally devastated world.

Jens Galschiøt's work has been a great inspiration for us in the completion of this book. He has generously shared his artwork for the illustrations, and on a visit by Anders Sandberg to the Galschiøt Gallery in June 2010 he also shared his time and ideas in conversation. We would like to thank Jens and all the staff at the Gallery for their generosity and hospitality. In the process of compiling this book, we are indebted to several other organizations and people. The Canadian Centre for Policy Alternatives has generously and enthusiastically supported the project from start to finish. Erika Shaker has edited all the chapters in spite of being on parental leave and then vacation. Nancy Reid has speedily put the manuscript in order in a hot and humid Toronto. Erika and Nancy have also provided valuable input on the title, the writing and the cover of the book.

The authors who ultimately made this book a reality have been wonderful in their response to hurried requests for submission and revisions of their chapters. The fact that this book took a mere three months to conceive of and produce is a testament to their dedication and compassion for one of the greatest environmental dilemmas of our time.

Anders Sandberg would like to thank his colleagues in the Faculty of Environmental Studies at York University for their support as well as York's Institute for Research and Innovation in Sustainability for working so hard to obtain observer status for him and Tor Sandberg at the Climate Change Conference in Copenhagen. He is also indebted to the undergraduate students and Teaching Assistants (some of whom have contributed chapters to the book) in "ENVS 1200: Taking Action: Engaging People and the Environment" who inspired the compilation of this book in the first place. They will now have a chance to read and discuss it.

Tor Sandberg would like to thank the alternative news media outlet rabble.ca which has inspired and assisted him in delving into the depths and crevices of climate change.

Finally we are grateful to Maria Legerstee, mother and wife, who is not only a Professor who edits her own books but also a capable and generous manager of our complicated and busy lives. Thank you Maria. Thank you mom. We love you.

L. Anders Sandberg and Tor Sandberg, Toronto, July 8, 2010.

Introduction

Climate change – who's carrying the burden?

L. ANDERS SANDBERG and TOR SANDBERG

Climate change refers to the increasing temperatures and changing weather patterns created by carbon and other greenhouse gas emissions and their negative consequences for humankind and the more the human world. Those who debate climate change often refer to it as global warming. The term global is telling. It suggests that climate change is a global phenomenon that is experienced by everybody. This is undeniably true, but it often follows that the same observers see the prescriptions as global too. An international community of experts, such as the Intergovernmental Panel on Climate Change, is considered to be a reliable source of information on the problem, and a collective of nation states, such as the United Nations Framework Convention on Climate Change, are considered the major agents for change.

The main strategic measures to combat climate change are often referred to as mitigation and adaptation. The two terms are clean, simple and seemingly value neutral. Nation states and their citizens have to mitigate climate change by reducing their carbon emissions. But they also have to adapt to climate change when the impact is already felt. Of the two, mitigation is clearly the main strategy advocated.

Mitigation through technological innovation, promoting electric cars, alternative energies, and energy efficiency, constitutes one of the major concrete measures to fight climate change. Another mitigation measure is the promotion of carbon markets, which includes concepts such as cap and trade and carbon offsets. Cap and trade is based on the idea that a progressively lower cap be set on carbon emissions within a specific area. Carbon emitters then trade emissions permits competitively in that space, giving them an incentive to innovate to reduce their emissions over time. Carbon offsets, on the other hand, allow companies and individuals to compensate for carbon emissions in different ways, such as the planting of trees or the establishment of an alleged green business in one place, which offsets a polluting behaviour in another place. The Norwegian Nobel Committee's choice of Al Gore for its Peace Prize in 2007 for his fight against climate change is a good example of the general acceptance of the green technologies and carbon markets options for dealing with climate change. Gore has made heavy investments in techno-fixes and carbon emissions exchanges in both the United States and Europe. He now makes handsome profits from these investments, a situation he is unapologetic about; after all, he claims, he is merely putting his money where his mouth is.

There are of course skeptics in the climate change debate. One of the most prominent is Bjørn Lomborg, the author of *The Skeptical Environmentalist* and *Cool It*. Lomborg is not a climate change denier; he concedes that climate change is occurring but that there is little anybody can do about it. He therefore advises that the money spent on climate change mitigation could be better spent elsewhere. Gore and Lomborg hold opposite positions on climate change mitigation measures (Lomborg, 2009). However, they are both market and technology enthusiasts who believe that the very economic structures, political systems, and technology networks that plunged the world into the climate change crisis in the first place, now have the ability to get us out of there. They thus obscure and marginalize the presence and possibilities of alternatives (Figure 1).

Climate justice advocates distance themselves from the prescriptions of Lomborg and Gore. Some of them, though, buy into the global discourse of markets in defining the climate change problem and its potential solutions. Most leaders of the developing

Figure 1: The climate change ruse. Former Vice President Al Gore and Bjørn Lomborg testify to the joint meeting of the Subcommittee on Energy and Air Quality and the Subcommittee on Energy and Environment, United States House of Representatives, March 21 2007. Though Lomborg and Gore hold contrasting views on climate change mitigation, they both advocate free market and technological solutions to the world's problems, be they climate change or world poverty (photo courtesy Bjørn Lomborg, www.lomborg.com).

world fall into this category and argue for massive transfers of financial and technological resources to compensate for the social and environmental costs incurred by their nations as a result of the climate change emanating from the massive carbon emissions in the Global North. Specific programs, such as the United Nations' Reducing Emissions from Deforestation and Forest Degradation program, or the Clean Development Mechanism of the Kyoto Agreement, fall into these measures. They either seek to put a market value on the ecological services provided by the world's rainforests as carbon sequesters (and then call for a commensurate financial compensation from First World nations to the Third World), or bring new clean industries to poor nations.

Some prominent individual climate justice advocates also appeal or access global institutions in an attempt to remedy climate change. Canadian Inuit leader Sheila Watt Cloutier, a First

Nations rights activist, has worked tirelessly through international institutions to address the climate change issue. Watt Cloutier was *co-nominated* with Al Gore by Norwegian Parliamentarians Børge Brende and Heidi Sørensen for the Nobel Peace Prize in 2007, though only Gore received the prize together with the Intergovernmental Panel on Climate Change. Stephen Lewis, in this book, represents a human rights advocate who is open to market mechanisms as one measure to fight climate change in light of its devastating effects on the poor countries of the world. Lewis' account of climate change in this book provides a dystopic and pessimistic view of the future, based on his work in Africa with the United Nations, human rights groups and local non-government organizations.

Many climate justice adherents, including most of the contributors to this book, are highly critical of the ability of modern global institutions to deal with the climate change situation. They join powerful critiques and add empirical support for the failures of the market mechanism and the techno-fix to deal with climate change (Gilbertson and Reyes, 2009; Leonard, 2010). They also unite with others who argue for local control of local resources, even where such control may mean leaving the oil in the soil (Angus, 2010; *Climate and Capitalism*).

Anders Sandberg and Tor Sandberg point to the content and bankruptcy of the official deliberations and proposals at COP15 in Copenhagen in December 2009, based on a first-hand account of their experiences at the Bella Centre, the official venue for COP15. They instead pin their hopes on the grassroots activists and transition movements at the people's Klimaforum and the streets of Copenhagen where participants worked to build different governance structures and livelihoods to create alternatives to the carbon economy.

Naomi Klein, also a participant at the Bella Centre and Klimaforum in Copenhagen, provides a powerful statement on climate debt, the debt owed to Third World countries for the devastation caused by carbon use and emissions in the developed world. Not only does she call for the developed countries to pay compensation to the Third World countries who are now unable to use the same carbon-development path, but she points to the necessity of drastic cuts in carbon exploitation and carbon emissions in the developed world. Klein, as elsewhere, calls for the formation of a grassroots transnational community to come together to pressure governments and world leaders to take action to deal both with social justice and climate change issues.

Vandana Shiva, in an interview with Tor Sandberg at the G8/G20 deliberations on financial debt in Toronto in June 2010, echoes the sentiments on climate debt by Klein. Shiva also provides telling connections between the climate change deliberations in Copenhagen and the formation of the G8. She argues that when the power of the United Nations Framework Convention on Climate Change and the many Third World nations represented on it threatened to put teeth in climate change policy, the initiative was thwarted by the power of the G8 and its involvement in climate change policy.

Sonja Killoran-McKibbin writes about the recent alternative people's conference on climate change in Cochabamba as a meeting of First Nations peoples and climate justice activists critical of the United Nations world summits on climate change. Initiated by the President of Bolivia, Evo Morales, the Cochabamba conference was all about building a new society that strives for extra-market solutions based on respect, reparations, and redistribution. But like its counterpart in Copenhagen, the delegates at the People's Conference also proposed global institutions to remedy the climate change crisis, among them the development of a Universal Declaration for the Rights of Mother Earth, a commitment of future COP meetings to tackle climate debt and climate refugees, a global people's referendum on climate change, and a Climate Justice Tribunal. Killoran-McKibbin notes the immense practical obstacles to conducting, let alone implementing, such global measures. In addition, she notes that the Bolivian government continues to depend on the revenue from extractive activities to finance social spending. She also comments that workers' conditions and regulations in the Bolivian mining sector remain more accommodating to mining capital than miners and the natural environment. Workers were in fact poorly represented at the Cochabamba conference.

Jacqueline Medalye elaborates on Killoran-McKibbin's point on the power of corporate interests in compromising the democratic pluralist assumption that all stakeholders have an equal voice in global institutions such as the United Nations Framework Convention on Climate Change. She instead points to the close connection between carbon business ventures and nation states, and the millions of dollars corporations spend on lobbying governments and other decision-makers to accommodate carbon-based industrial growth. She also points to the increasing police presence at COP15 which served to both exclude and punish the environmentalist groups at the conference.

Aaron Saad concludes the first part of the book with a chapter on climate refugees, those growing number of people who are impacted by climate change (and related factors) to the point where they feel compelled or are forced to move elsewhere. Saad notes that the several attempts on the part of concerned individuals and organizations to include climate refugees in international agreements and conventions have been repeatedly scuttled. He also points to the iniquitous historical foundations of climate change in colonialism but concludes that raising such issues typically falls on deaf ears in international negotiations.

In the first part of this book, the contributors take climate change and climate justice and the various deliberations surrounding and radiating from them at the recent global venues in Copenhagen and Cochabamba as their starting point. The chapters, whether they are more or less critical of the negotiations at these conferences, suggest that the pressures and alternatives articulated by climate justice advocates can make a difference. But there is also a profound sense of pessimism permeating these deliberations given the continued political and economic power of the large oil corporations and the support they receive from national governments both financially and in suppressing dissent.

In the second part of the book, we change the focus from the global prism of Copenhagen and Cochabamba to the conceptual and local levels of climate change. We challenge the monopoly of corporations, nation states, experts and environmentalists in defining, identifying the terms for discussion, and framing of the solutions to climate change. We turn to some of the communities that suffer the most from the impacts of climate change. Many communities, we argue, live in the chilly climates created by the various discourses, living patterns and ideologies that go along with climate change. Feminists originally coined the term chilly climate to describe the chilly reception women have faced in historically male-dominated occupations. But the concept can also convey a situation where an individual or various groups face discrimination or unfair treatment or where alternatives positions and ways of knowing, thinking and living are marginalized by the climate change dilemma.

Neil Adger et al. (2001) have argued that there are two general discourses in the climate change debate. One they call a global environmental management discourse which represents the technofix and market solutions referred to in the above. The other they refer to as a populist profligacy discourse. The profligacy discourse refers to the "over-consumption" of resources in the Global North as the problem and solution to the climate change situation. It also portrays local actors as victims of external interventions that bring about degradation and exploitation. Adger et al. (2001) argue, however, that aspects of vulnerability and adaptability are made illegible in the profligacy discourse. On the one hand, the restructuring of the ownership patterns and the control of resources in a particular locale may cause the area to be more vulnerable to climate change. On the other hand, a change in the local social structure may lead to a collective effort to rehabilitate the local environment to make it more resistant to climate change. There are thus direct connections between climate change and seemingly separate political, social, and economic conditions.

One group of climate justice advocates and analysts focus precisely on those that suffer the most from climate change, both at the points of extraction of the carbon, and at the sites where the impact of carbon emissions are felt the most. These include communities such as First Nations living near the Tar Sands development in Alberta (Clarke, 2008), the Indigenous people expelled from their oil-rich lands in Iraq, and the people and First Nations residing in the midst of the refineries in Sarnia, all who pay a high price socially and environmentally.

In their telling title, "Framing Problems, Finding Solutions", Stephanie Rutherford and Jocelyn Thorpe argue that the way we frame problems determines their perceived solutions. They note that while the Canadian federal government has taken some responsibility for the problems experienced by Canada's First Nations people, it has not acknowledged fully the appropriation of First Nations' lands. Instead, the government embraces the concept of wilderness as worthy of protection (nature and national parks), though such concepts are social constructions that erase First Nations' perceptions of the very same areas as home. Similarly, they argue, problems and alleged solutions to climate change are seen as universal, and First Nations lands are often seen as a prescription (as carbon sinks, for example), while First Nations visions and claims to the very same lands are obscured.

Noël Sturgeon also calls attention to how dominant popular stories about climate change normalize the very social structures that are part of it. In North America, the so-called nuclear family, based on the heterosexual union between a man and a woman, both normalizes the destructive suburban consumer society that is responsible for climate change while pathologising and marginalizing other forms of family formations that may tread more lightly on the Earth. Sturgeon shows that popular culture is often solicited by different positions on what constitutes normal family values. Fundamentalist Christians, based on the award-winning documentary film *The March of the Penguins*, have claimed penguins as the "normal" and archetypical representation of the nuclear heterosexual family, while the gay community, based on the frequent occurrence of homosexuality among penguins, has done the exact opposite, arguing that penguins represent the normality of gay marriage. Penguins, on the other hand, refuse to stick to any category. Sturgeon draws on the different penguin family values to distinguish between reproductive rights and reproductive justice. The former recognize the rights of women to determine the fate of their bodies (though legal and technological means), while the latter expands the concept of reproductive rights to all people (and sometimes animals) to form different family, productive, consumptive and other forms of unions.

Jelena Vesic elaborates on the theme of inclusion and exclusion in the climate change debate. She shows how the Ice Bear Project by sculptor Mark Coreth, sponsored by WWF World Wide Fund for Nature and Polar Bears International at COP15 in Copenhagen, idealizes, romanticizes and puts into prime focus the polar bear as a victim and point of action against climate change. This image neglects the resiliency of both polar bears and Inuit people. She shows how polar bear management since the 1960s has contributed to an increase in the population; that some populations are stable in spite of climate change; that the hunt, even the trophy hunt, which is led by the Inuit, is respectful of the bear and contributes to the cultural survival of the

Inuit; and that most natural scientists are in favour of the hunt. Such a partnership between the Inuit and the polar bear that has lasted for generations in the North is obscured and marginalized by the construction of the polar bear as the symbol of the fight against global climate change.

Isaac 'Asume' Osuoka provides an analysis of Operation Climate Change in the Niger Delta of Nigeria where local people have fought the destruction and social devastation resulting from oil exploration and exploitation for decades. Operation Climate Change is a telling term for the protest movement in the Niger Delta, because it reveals the interconnection between extraction and emissions of carbon. Osuoka takes the reader on a toxic tour of the Niger Delta showing how the crude exploitation of oil for export, combined with a rusty and corroding infrastructure and the flaring of natural gas (considered uneconomical to use), results in a polluted environment. He also brings into stark relief the state violence that is used to maintain the carbon economy. The situation in the Delta is clearly and disturbingly connected to colonial exploits, both historical as well as current.

Tanya Gulliver looks at the devastation caused by Hurricane Katrina, seen by some as emblematic of the consequences of climate change, as more a function of the vulnerabilities built up in New Orleans where the natural (wetlands) and human-made (levees) barriers to hurricanes have been degraded, and a human population, primarily Black and/or poor, has suffered the effects of Katrina to an extent greater than others. As in the Niger Delta, this is because of the vulnerabilities built up in Louisiana as result of the practices of the petro-chemical industry in the area. Gulliver's chapter, echoing Osuoka, shows the linkages and inter-relationships between the environmental devastation in both time and place, between Louisiana and the Niger Delta, and between the long-term environmental pollution in New Orleans and the Gulf of Guinea, and the sudden and drastic effects of the 2010 oil spill in the Gulf of Mexico. The very same corporations have been active in the same areas over a long period of time.

Jay Pitter suggests that it is not only the voices of subjugated communities, such as those of the Inuit, Ogoni, and New Orleanians that can and should be given voice, but also those of individuals within or connected to such communities. Pitter shows how a casual inquiry inviting friends and family to a conversation about environmental narratives can lead to powerful insights. Her chapter uses a narrative tracing the travels of young woman and her mother from Toronto to Africa as a touchstone in the process of unearthing a subjugated narrative grounded in the environmental and social conditions of everyday life. Pitter's account invites all of us to conduct similar inquiries with people in our immediate environments.

Personal stories and social justice concerns and positions are often marginalized, obscured and forgotten in the chilly climates that surround the climate change discourse. The contributors to Part 2 of the book insist that climate change needs to be seen in context, as part of a larger system that despoils and exploits, and that contains different vantage points and perspectives on the lived experiences and imagined solutions for atmospheric climate change. They suggest that imaginary thinking, alternative solutions, and other voices are crowded out in the chilly climates of dominant discourses.

In the third part of the book, we explore some of the experiences and actions that directly challenge the carbon economy and its associated capitalist economy to propose alternatives or at least the elements of alternatives and positive actions for change.

Elizabeth May provides an account of how environmental education curricula fail the current young and future generations by passing on environmental stewardship responsibilities to them, adopting environmental lesson plans from corporations, and paralyzing students with fear. She argues for a more inclusive and socially aware environmental agenda that incorporates green issues in all disciplines of educational curricula, reacquaints students with the non-human world through experiential learning, and builds active environmental citizens that are engaged in community actions.

First Nations peoples, as the Cochabamba conference showed, are often the first in line to bear the burden of climate change. This is certainly true of the First Nations in Canada, whose living standards and general health are well below that of the Canadian average. Lakhani et al. point in particular to the high HIV/AIDS epidemic among First Nations youth in Canada, and describes an action to use of hip hop to fight the disease. Hip hop, which originated among Black Americans, they show, has become the voice of oppressed youth globally, and constitutes one means to de-colonize the chilly climate that surrounds climate change discourse and development.

Anders Lund Hansen explores the largest and longest standing squatter settlement in Europe, Christiania in Copenhagen, as an inspiration for alternative living. Christiania, Lund Hansen suggests, is facing enormous challenges in Denmark where a right-wing government is seeking to promote or 'normalize' (what Sturgeon call 'naturalize') a business-friendly urban environment where private property reigns supreme in the land and housing markets. Christianites, by contrast, have lived under 40 years of communal ownership of housing and land, as well as promoted a life of accommodation of difference, which has included a less carbon-intensive and less polluting urban environment. Lund Hansen recalls one of many struggles where Christianites have successfully fought the state to maintain their alternative vision of society.

Adrina Bardekjian Ambrosii investigates the Transition Town movements and Climate Camps in the United Kingdom as opposition strategies and alternatives to the carbon economy. She argues that both the Transition Towns, which are long-term reformist attempts to reduce carbon dependence within existing governance structures, and the short-term protest Climate Camps that call attention to particularly dirty carbon emitters, and challenging the fundaments of capitalism, contain valuable lessons and personal stories in the overall scheme to move beyond climate to system change.

Deborah Barndt concludes the book by pointing to an initiative in the Greater Toronto Area that seeks to build so-called foodsheds that integrate local sustainable food production with local consumers, keeping in mind social justice issues that make food available to all. Calling her essay "Digging Where You Stand," Barndt emphasizes the importance of building local networks that include the university, non-government organizations, farmers and local residents to build food systems that are socially equitable, environmentally sustainable and civically engaged.

The contributors to this book, then, first point to the entrenched problems associated with the current dominant strategies to deal with climate change. They acknowledge the devastating effects of climate change and the failure of modernist institutions grounded in the market, the technological fix, expertise, the nation state, and the international negotiation framework. They instead point to several politics at different scales that provide alternative ways of seeing and acting. These include the pursuit of what Saad calls global social justice, putting continued pressure on conventional modernist institutions to act, respond, and modify their policies along climate justice lines. But they also direct us to re-conceptualize and re-localize the climate change debate so that it focuses on those who suffer the most from climate change, yet may hold the key to powerful alternative ways of thinking and acting. Finally, the contributors urge support for grassroots initiatives, be they in the classrooms of Canadian elementary schools, First Nations communities in Ontario and the North, the freetown of Christiania in Copenhagen, the Transition Towns and Climate Camps in the United Kingdom, or the community-formed foodsheds promoting organic and accessible foods in the Greater Toronto Area.

* * *

REFERENCES

Adger, W. Neil; Benjaminsen, T.; Brown. K.; and Svarstad, H., 2001. "Advancing a Poltitical Ecology of Global Environmental Discourses," *Development and Change*, 32, 681-715.

Angus, Ian, editor, 2010. *The Global Fight for Climate Justice: Anticapitalist Responses to Global Warming and Environmental Destruction*. London: Resistance Books.

Bond, Patrick, 2010. "Circumventing the Climate Cul-de-Sac: Charleton-Cochabamba-Caracas vs. Kyoto-Copenhagen-Cancun," *Social Text: Periscope*, 28 March. http://www.socialtextjournal.org/periscope/2010/03/swap.php, accessed 5 July 2010.

Clarke, Tony, 2008. *Tar Sands Showdown*. Toronto: James Lorimer.

Climate and Capitalism, various issues.

Gilbertson, Tamara and Reyes, Oscar 2009. "Carbon Trading: How it works and why it fails, *Critical Currents*," no. 7. Uppsala: Dag Hammarskjold Foundation.

Lomborg, Bjørn, 2009. "Mr. Gore, Your Solution to Global Warming is Wrong," *Esquire* (August). http://www.esquire.com/features/new-solutions-to-global-warming-0809?click=main_sr, accessed 21 February 2010.

Leonard, Annie, 2010. The Story of Cap and Trade. http://www.storyofstuff.com/capandtrade/, accessed 29 May 2010.

Part I

Climate Change and Climate Justice

STEPHEN LEWIS

The Health Impact of Global Climate Change

The following is an edited version of Stephen Lewis' keynote address to the
5th World Environmental Education Congress, May 10, 2009, in Montreal.

[...]

In 1988, I was fortunate enough to chair the first major international conference on climate change. We had between three and four hundred scientists and politicians gathered together over several days. The debate was of enormous intensity and at the end of it, a declaration was drafted, the opening words of which read as follows:

> Humanity is conducting an unintended, uncontrolled, globally pervasive experiment whose ultimate consequences could be second only to a global nuclear war. The earth's atmosphere is being changed at an unprecedented rate by pollutants resulting from human activities, inefficient and wasteful fossil fuel use, and the effects of rapid population growth in many regions. These changes represent a major threat to international security and are already having harmful consequences over many parts of the globe. (World Conference on the World's Atmosphere in Toronto, 1988)

More than 20 years later, that opening portion of the declaration is an adequate and entirely legitimate representation of the way people feel about the onset of climate change and global warming today. More than 20 years ago, the aspects of the impacts of global climate change were itemized thus:

These changes will imperil human health and well-being; diminish global food security through increases in soil erosion and greater shifts in uncertainties of agricultural production, particularly for many vulnerable regions; change the distribution of seasonal availability of fresh water resources; increase political instability and the potential for international conflicts; jeopardize prospects for sustainable development and reduction of poverty; accelerate the extinction of animal and plant species upon which species survival depends; alter the yield productivity and biological diversity of managed and ecosystems, particularly forests. (ibid.)

This itemization of the impact of global warming has been authenticated time and time again over the intervening 20-plus years. We have done very little, I point out, to address those problems in a way that can be seen to confront climate change and to reverse the consequences.

Interestingly, the most vivid moment of the conference occurred when the minister of the environment for Indonesia took the platform. Emil Salim was a very gentle and sweet man, and he looked out at the audience and said: "If you think that Indonesia is going to curtail its economic growth in order to compensate for the environmental degradations of the western world, you're crazy. And unless you transfer the necessary technology to Indonesia, unless you give us the resources so that we can shift from our reliance on fossil fuels to alternative energy sources, we're not prepared to accommodate your demands." And that became the issue of the conference, and is precisely the issue today.

[...]

That brings me, if I may, to a personal view. I am now in my dotage for heaven sakes. I'm limping into senility, I'm 71 years old, I no longer observe any of the diplomatic proprieties. I'm going to speak to you from the heart and as honestly as I can.

In order to avert the crisis that is looming, we have to create global citizens. We have to create citizens with acute environmental sensibilities, with a profound and honest understanding of the issues at stake. In their preface to one of the volumes on education, UNESCO writes as follows:

This is an era in which some one billion people live in poverty while the majority of the world's wealth is in the hands of just a few people. This is a time of considerable turbulence and instability — a time of financial and economic crisis, and social upheaval as well as persistent ecological degradation, global warming and the rampant consumption of finite resources. As the current crisis is likely to affect everyone, it is time to anticipate possibilities for profound transformation, toward more inclusive societies, more equitable growth, and more responsible behaviours of consumption. (UNESCO, 2009)

As I look at the advent and enormous propulsion of the process of global warming and climate change, three things come to mind.

The first point: the bulk of environmental education today tends to address the greening of society. Whether it's suggestions from Ontario's Ministry of the Environment, or whether it's the greening initiatives set out in the various UNESCO documents, everybody talks about recycling, using the right light bulbs, driving hybrid cars, taking shorter showers, travelling on public transit, building solar panels, planting trees or setting up wind power. All the apparatus of environmental education focuses on the importance of greening society and becoming more aware and more sensitive to environmental priorities. It is tremendously admirable for its consciousness-raising and for creating a more liveable community, both locally and internationally.

But it's not sufficient. And it seems to me that it's almost an exercise in irresponsibility to pretend that having kids go out into the countryside, and dwell beside the water, and fish in the ponds, and plant trees, and see the beauty of nature — it's not enough. It's not nearly enough because, in truth, all of that will not prevent the growing crisis of global warming.

That leads me to the second point I want to make. The only answer to this crisis is the most dramatic reduction in the dependency on fossil fuel and the discharge of carbon; everything else is incidental. We are not going to rescue the planet with environmental education that focuses on greening possibilities — without at the same time acknowledging that we're in a tremendous race against time. What is involved here is the need to focus young people, who are being educated at whatever level, on the reality that unless we deal with the discharge of carbon, we are dooming the planet. This isn't some abstraction.

What has happened between 2007-2009 is absolutely catastrophic in its implications. All of the projections, all of the predictions of the Intergovernmental Panel on Climate Change are proving to have been wrong. They understood that we were in a precarious position, but they never fully understood the rapidity of the changes that are occurring, particularly the changes that are occurring with the melting of the polar caps.

These changes simply unsettle everyone — and it seems to me that we must be prepared to engage in environmental education that says: We're in a crisis unlike any other; we're facing the possible catastrophic implosion of humankind; we're going to have an apocalyptic event somewhere between 2030-2050. It may be 30 million Bangladeshi environmental refugees as a result of rising sea levels and the inundation of coastal sea regions, or it may be an incredible calamity at the southern end of the African continent where there isn't enough food to feed people. But the truth of the matter is that we have unleashed forces which are not being curtailed, and everybody recognizes that what is required is political will to reverse the process.

One of the co-chairs of the 5th World Environmental Congress, Lucie Sauvé, writes in the introduction to the conference program: "In the wake of preceding environmental education congresses, this 5th Congress becomes a political act." I repeat:

> This 5th Congress becomes a political act. Education and the environment have a strong political dimension. This congress aims to contribute to the recognition that the socio-political importance of environmental education and to strengthen this field with the support of the decision makers.

And there was a remarkable piece of writing just a few months ago in the *London Review of Books* by John Lanchester, please forgive my quoting, but I think it's important:

The remarkable thing is that most of the things we need to do to prevent climate change are clear in their outline, even though one can argue over details. We need to insulate our houses, on a massive scale; find an effective form of taxing the output of carbon (rather than just giving tradeable credits to the largest polluters, which is what the EU did — a policy that amounted to a 30 billion euro grant to the continent's biggest polluters); spend a fortune on both building and researching renewable energy and DC power; spend another fortune on nuclear power; double or treble our spending on public transport; do everything possible to curb the growth of air travel; and investigate what we need to do to defend ourselves if the sea rises, or if food imports collapse. (Lanchester, 2007)

We know all this, but whether it will actually happen is a different question. There is simply not the political will yet in evidence. In Canada, with the exception of some initiatives in some provincial jurisdictions like Quebec and British Columbia, and to a much lesser degree Ontario, it's simply not happening.

Canada is becoming one of the grossest international violators of the prescriptions for climate change. We have a federal government (forgive me for being political for a moment; I'm just going to have an ideological spasm, and then attempt to contain myself) which refuses to react to the reality of global warming, and permits an increase in the discharge of carbon that accelerates with every passing year. Even though Canada pretended to be a part of Kyoto 1, and signed on to reduce emissions by 6%, instead emissions are up 26%. It is a travesty, but it's also, historians will say, a matter of criminal negligence because the way the world is moving is a nightmare. We simply have to find an alternative way, whether it's a carbon tax, whether it's cap and trade, whether it's sophisticated measures of sequestration, whether it's the tremendous embrace of the alternative renewable energy, solar and wind in particular. However much nuclear power unsettles people, the Swedes, the French and the British are now looking at the nuclear option. All of this has to be reviewed. Why? That leads me to the third element that I want to explore briefly.

It may be too late. I want to be honest. I think it is too late. Sir Nicolas Stern said that there's nothing we can do now that will influence what will happen between 2030-2050 because climatic factors are in process which are irreversible and cannot now be changed. I think that environmental education now consists in large measure of explaining to the students that we're heading for catastrophe and that we have to find unusual responses in order to address that catastrophe.

Isn't it interesting that last week in *Nature* magazine, the debate was intensified when a group of scientists concluded the world had little chance of holding a temperature rise of two degrees Centigrade — a level widely regarded by scientists as the limit of safety beyond which climate change becomes irreversible and potentially cataclysmic.

The Financial Times, which I think you'll agree with me is hardly a left-wing rag, had a most remarkable article called "Changing the planet might help preserve it," in which they begin to explore approaches to dealing with global warming which were originally thought to be completely off the wall, completely scatterbrained, geo-engineering solutions that no one would take seriously. Listen to the way this article starts:

> A giant mirror drifts slowly through space between the Earth's surface and the sun, intercepting the rays of sunlight before they reach the Earth, and deflecting them safely away. The mirror, made up of millions of silicon chips, is situated at a point in space where the sun's gravity and the Earth's cancel each other. This vast structure, assembled painstakingly for years by spacecraft, drifts naturally away from its starting point over time, but complex on-board systems nudge it gradually back to resume its vital role in keeping us safe. This space mirror is — so far — science fiction. Such a structure would cost hundreds of billions of dollars, even if it were technically feasible. But soon many scientists say we may need to start building space mirrors, creating artificial clouds or altering the chemistry of the sea to prevent the worst effects of global warming. (Harvey, 2009)

Suddenly all of the science fiction scenarios become potential pragmatic responses to a looming cataclysm, and therefore it seems to me that there is nothing more important than environmental education. I mean, in a way, it's a privilege to be at the heart of an intimate struggle internationally over the preservation of the planet. But the very real question is: can we reverse it now? Have we unleashed Armageddon? Has it gone too far to be coped with? Must we now begin to contemplate measures of geo-engineering which were formerly thought to be faintly lunatic in their scientific prescriptions?

There are a large number of very credible scientists who think that geo-engineering will inevitably be the alternative on which we'll have to rely, because it is clear that the world will not diminish its dependence on fossil fuels. It is clear that no matter what we contrive, China will continue to build one coal-fired power plant every week, with the dirtiest coal discharging the highest measure of carbon in the world. It is equally clear that even with financial turbulence, we're going to be building more and more runways and airports and ever larger planes and there is no fuel in sight that will diminish the consequences of carbon discharge from flight.

It simply becomes clear that even with the best will in the world, Kyoto 2 may not be enough to reverse what is underway. And therefore environmental education has to explore every one of these aspects. It's thrilling in one sense because it's such an extraordinary intellectual exercise; it's rather depressing in another sense because it's a commentary on how paralyzed the international community has been in its response.

In 1987, Gro Harlem Brundtland warned us (in the Commission on Environment and Development), and UNEP has warned us time and time again about what we're facing. We have failed to pay serious attention to it and now we may have inherited the whirlwind. Along the way there are going to be some grievous consequences for humankind.

Let me briefly enumerate them for you: the shifting agricultural patterns, as a result of global warming, will make life virtually unbearable in many parts of the planet. In particular, the Intergovernmental Panel on Climate Change indicated in its most recent report that southern Africa would feel most forcefully the consequences of global warming.

It's like some kind of conspiracy. All of these countries in southern Africa are wrestling desperately with HIV/AIDS. They have prevalence rates that are a nightmare. The pressures on these countries are overwhelming. I spend a lot of my time in Africa. You can't imagine how people struggle with such resilience and courage to keep their lives and communities together. One of the truths about Africa is the tremendous sophistication, intelligence, generosity of spirit, and fundamental human decency evident at the grassroots, at the community level. Then on top of all the poverty, where more than half the population lives on less than a dollar a day, and on top of the pandemics of AIDS and tuberculosis and malaria, Africans are likely to experience ever-more severe droughts. There will inevitably be a reduction of agricultural productivity, a reduction of household food security, more and more famine, more and more hunger, and the possibility of conflicts over water. It's a nightmare.

I read the compendium of news stories that come across my computer screen every morning, and I saw this, back in January, from the Science and Development Network: "billions face food shortage this century, warn study. Harvests of maize, and other staple crops could drop by up to forty per cent by the end of the century." Forty per cent: the staples on which people depend! I'll be long gone, but our kids and our grandchildren, how in God's name are they going to cope with the apocalyptic dimensions of what is occurring? Just last month, in the *Times* in the United Kingdom, there was a story: "billions of people face famine by mid-century, says top U.S. scientists." Let me simply read it to you:

> Famines affecting a billion people will threaten global food security during the 21st century, according to a leading U.S. scientist. Nina Fedoroff, the U.S. State Department chief scientist, is convinced that food shortages will be the biggest challenge facing the world as temperatures and population levels rise. (Smith, 2009)

What are we doing in this world? Why are already vulnerable populations thus expendable? Where is the moral anchor of the international community? How have we allowed this to go on this far? I spend a lot of time in countries where people are already impoverished. The soaring food prices have put another hundred million people below the poverty line of a dollar a day. It's beyond conscience to imagine that our passivity and negligence on global warming will result in further compromising so many millions of lives on another continent.

In the case of water, the World Health Organization points out that roughly 1.5 billion people back in 1990 will increase to 3.6 billion people in 2050 living under water-strained situations. Around air pollution the WHO says that extreme high temperatures can kill directly. It has been estimated that over 70,000 excess deaths were recorded in the extreme heat summer of 2003 in Europe. With respect to disease transmission, the mosquitoes who carry malaria are moving into countries that never had malaria before as warming patterns accelerate. And one of the things that absolutely throttled me, while I was reading the World Health Organization's analysis, was their reference to the probable acceleration of sexual violence.

When societies are destabilized by disease or poverty or acute hunger, when there is a tremendous amount of internal turbulence, then that turbulence is often expressed through violence, and increasingly sexual violence. Whether we're talking about the Khmer Rouge in Cambodia, or the crushing of independence in East Timor where Indonesian government troops engaged in a massive campaign of raping and sexual violence. And I remind you that the military leaders in the Balkans, white western countries, are now serving time for crimes against humanity rooted in rape. This pattern of physical and sexual violence is likely to career right out of control when there will be destabilization as a result of climate change and global warming.

During the electoral violence in Zimbabwe, Mugabe unleashed the so-called war veterans and youth core to engage in a campaign of politically orchestrated sexual violence. If you worked for the opposition, and you were a woman, you were raped and tortured.

I'm part of a little NGO in the United States which is taking affidavits from the women who were subjected to sexual violence in Zimbabwe. The whole world knows what happened and refuses to do anything about protecting women from the marauding militias. And that, let me tell you, is a manifestation of gender inequality. Anything in the world that accentuates this inequality is so profoundly wrong that it should rally all of humankind. The most important struggle on this planet is the struggle for gender equality. You cannot continue to marginalise 52% of the world's population and ever expect to achieve social justice or equality; it's just not going to happen.

So, the reality is that the phenomenon of global warming is inciting a myriad of implications and complications and this is, or ought to be, the stuff of environmental education.

It is absolutely unbearable that young people are going to have to live with the consequences that we have created. I've often thought, in my own life, that I should have spent a lot more time working on environmental issues. I feel a kind of insensate guilt and shame that 20 years ago I was part of a conference that forecast what was coming, and I chose to do other things and find other priorities in life.

I have two grandsons of seven years of age and four years of age and I can't stand the thought of what they're going to inherit. But, you are collectively at the heart of it. You can conceivably turn around an apocalypse. I'm not sure it's possible, but if it is possible, it will come through environmental education, it will come through your collective, skilful, principled, uncompromising, leadership.

You are extraordinarily privileged to be an environmental educator at this moment in time. I salute you for it.

* * *

Stephen Lewis is a Professor in Global Health at McMaster University and Chair of the board of the Stephen Lewis Foundation. Among several senior UN roles that spanned over two decades, Mr. Lewis was the UN Secretary-General's Special Envoy for HIV/AIDS in Africa; Deputy Executive Director of UNICEF; and Canada's Ambassador to the United Nations.

REFERENCES

Antony, Naomi, 2009. "Billions face food shortages this century, warns study," *Science and Development Network*, January 9. http://www.scidev.net/en/news/billions-face-food-shortages-this-century-warns-st.html.

Harvey, Fiona, 2009. "Changing the planet might help preserve it," *Financial Times*, May, 8. http://www.ft.com/cms/s/0/121f650e-3bea-11de-acbc-00144feabdc0.html.

Lanchester, John, 2007. "Warmer, Warmer," *London Review of Books*, March. http://www.lrb.co.uk/v29/n06/lanc01_.html.

Smith, Lewis, 2009. "Billion people face famine by mid-century, says top U.S. scientist," *Times Online*, March 23. http://www.timesonline.co.uk/tol/news/environment/article5962238.ece.

UNESCO. 2009. *Good Practices Education for Sustainable Development*. The United Nations Educational, Scientific and Cultural Organization. http://unesdoc.unesco.org/images/0018/001812/181270e.pdf.

World Conference on the Changing Atmosphere in Toronto, 1988. *The Changing Atmosphere: Implications for Global Security*. http://www.cmos.ca/ChangingAtmosphere1988e.pdf.

**L. ANDERS SANDBERG and
TOR SANDBERG**

From Climate Change to Climate Justice in Copenhagen

The results of the fifteenth United Nations Conference on climate change in Copenhagen (COP15) have been variously described as abysmal to farcical. Even sympathetic observers concede that the "agreements" will do little to stall the growth in global carbon and other greenhouse gas emissions and their deleterious effects on human and natural environments. The record of the conference in fact challenges the world community to think about climate change in different ways. We found this out on our trip to COP15. Representing York University and the alternative on-line media organization rabble.ca, we were starkly confronted with two distinct stories about climate change.

The dominant story informs the government negotiators and campaigners that work into the night to find a global agreement. It assigns trust and hope to modern institutions such as the nation state, the corporation, the market, technological fixes, the natural science community and its funding agencies, and concerned publics and environmental organizations lobbying national governments and international organizations in reversing the current trend in carbon emissions. Negotiations centre on cap and trade, carbon offsets, carbon capture and storage, market and techno-fixes, and so-called global agreements beyond Kyoto.

We found an alternative story at the conference articulated by critics of the dominant narrative and the many transition movements that advocate for radical challenges to the forces that profit from the continued exploration and extraction of fossil fuels. These movements also see their participation in international deliberations as undemocratic and restrictive. This story trusts humanity's ability to cope with climate change but doesn't believe modern institutions can do so. Critics of the dominant narrative insist on the re-emergence of diverse local to global networks of co-operative need-meeting structures. They also focus on the skewed distributional benefits and costs from climate change and advocate for a massive climate debt owed Third World countries for bearing many of these costs. Their story sees climate change as a symptom of a problem; it therefore calls for systemic change. The rallying cry for the transition movements is climate justice.

11 / 12

We arrived in Copenhagen on Friday, December 11, on the fifth day of the COP15 Climate Change Conference. We went straight to the Bella Center, the main venue where the dominant story played itself out through deliberations by the official delegates of nation states, who were there to hammer out a deal on the restriction of carbon emissions.

Issuing observer status to various outside groups is common for United Nations conferences, and the Copenhagen one was no exception as organizers issued three times more permits than the Bella Center could accommodate.

But obtaining such permits was not easy. We had received official non-governmental observer status through York University, but only after a request by the United Nations that the President of York give his written endorsement. As we arrived we had to go through a baggage and body search before entering and had to, as well, present confirmation of our status. But the registration went really well. We waited no more than half an hour in line and got our papers and photos taken for the admission identity card. We also got a transportation pass that allowed us to travel free on all trains and buses in the greater Copenhagen area and the province of Skåne in Sweden. On our very first day in Copenhagen, we were well accommodated and had easy access to the official deliberations.

Once inside the venue we discovered that the Bella Center was also a place for a broader public discourse and exchange; the place was packed with environmental organizations, ranging from those that peddled various environmental technologies to those that promoted environmental justice in different parts of the world. Jet-lagged and tired, we managed to take in one late afternoon session at the Bella Center featuring Bill McKibben, founder of 350.org; Cape Verde Ambassador Antonio Lima (whose island state is drowning due to rising sea levels); and Ricken Patel, Executive Director of Avaaz — a global progressive web movement. We received our first taste of reformist action at the meeting, environmentalist organization and nation state representatives working together to call attention to the small island states that face the imminent impacts of climate change. The panel worked within conventional frames, seeing restitution and amends through the institutions of the nation state (as represented by the ambassador) and the international negotiation process. The panel was backed by a gallery of grassroots activists who performed acts of support in front of a tired group of media people.

Figure 1: Side event at the Bella Center (photo by authors).

12 / 12

As we walked through Copenhagen on our second day at the conference, we were confronted with the techno-fixes and reforms of capitalism that were seen and marketed through the city as solutions to climate change. The progressive steps of ecological modernization in the shape of alternative means of energy production and transportation were visible everywhere. We were reminded of this pattern when looking out over Öresund, where the wind turbines hummed, and when seeing the stream of bikers zipping by the stalled cars in the streets. Yet such measures have not made a significant dent in the increase in carbon emissions.

We also saw a multitude of corporate advertisements that espoused care about climate change and a keen interest for solutions. Sweden's construction giant Skanska and real estate firm Norrporten simply declared "Stop climate change. Make Cop15 matter" (Figure 2). Both Coca Cola and the German giant Siemens referred to Copenhagen as Hopenhagen, a place where energy efficiency and modernization combine to stave off the negative effects of climate change.

Figure 2: Co-opting the opposition. An "aggressive" corporate response to climate change by Skanska, one of the world's largest construction companies, and Norrporten, one of Sweden's largest real estate companies (photo by authors).

Figure 3: Siemens advertisement on bus shelters (photo by authors).

But the glossy corporate messages exhibited on posters were not uncontested. In one Coke advertisement adbusters had added the words: "Dear Coke, it's great you're doing something good, but it doesn't cancel out all the dangerous s*it you do." In a Siemens ad, the company's name had been partially covered by a sticker reading: "System Change, Not Climate Change." The scribbles and stickers on the advertisements reminded us that there are those that see climate change as a symptom of something much more fundamental, the increased corporatization of the global economy, and its control of a small group of powerful interests that are active across the globe.

The adbusters prepared us well for our next destination, Klimaforum 09 — The People's Forum on Climate Change, which represented the climate justice storyline. It was buzzing with activists who were highly skeptical of and disillusioned with most of the official efforts at the Bella Center. They questioned the contention that further economic growth of gas and oil extraction and lower carbon emissions were at all compatible. But most profoundly — the recurring theme permeating the forum — was a call to build networks, meshworks or transcommunities within and across societies from the local to the international scale. The message is that the global is not enough, nor is the local, but we need to scale efforts up and down to come to viable solutions.

We felt more at home at the Klimaforum, though we were certainly not given a consistent view of the alternative position that was free from contradictions. We first attended a session organ-

Figure 4: Klimaforum 2009, The People's Climate Forum, was open to anybody who chose to attend. No security passes, admission tickets or body searches were required (photo by authors).

ized by a Belgian union-funded Christian coalition assisting workers in retrofitting their homes. It did so by forming purchasing coalitions for both insulating material and heating oil. In Belgium, where home ownership is high even among the poorer segments of the population, it was an important reminder that not everyone can afford to retrofit their home with solar panels and windmills. A Dutchman got the same message after he challenged the coalition for working toward reducing the price of oil. 'Higher prices lead to decreased consumption,' the Dutchman argued, without considering the effect on the poor. With such oversight, it's no wonder the climate change movement hasn't resonated as profoundly in poorer communities in the industrialized North. The larger lesson that climate change reformers have to consider is building coalitions with working and poor people for whom carbon fuels may still play an important role.

In the afternoon we linked up with a group of academic activists from Scotland, Wales and Scandinavia for an "Academic Seminar Blockade". They were also disillusioned with the conventional avenues to tackle climate change, and gathered to present papers on the topic "Climate Change: Power, Policy and Public Action." Kelvin Mason from the Graduate School of the Environment and the Centre for Alternative Technology at Machynlleth, Wales, led the planning of the Blockade session, outlining some of the recent alternative initiatives challenging climate change in the United Kingdom. The Transition Town movement contains both promises and lessons. Initiated by Rob

Hopkins, in Totnes, Devon, the movement now includes about 60 towns in the United Kingdom and strives to use permaculture to promote local self-sufficiency in foods and energy, as well as build community solidarities. Critics have charged that the movement is naïve, reformist, and does not address the fundamental power structures that underpin a capitalist economy. They therefore argue for a more radical approach, suggesting there is a rocky road to a real transition.

Climate Camp campaigns take notions of power more seriously than the transition town movement advocates. They typically set up camps as squats where folks form small groups that cook, live sustainably, learn together and plan for civic actions that target a specific project, such as an airport, coal fired energy plant, highway or rail line. Such direct action signals a rejection of such conventional institutions, and call for alternative societies, means of transportation, and day-to-day living.

A consensus seemed to suggest that activists be sensitive to specific situations. Routinely, for example, activists who arrive to at a protest site reject car travel and opt for a vegan diet while local residents see such practices as strange or impractical. One speaker talked about the initiatives of Welsh farmers initiating green energy ventures in light of their struggles emanating from the cost-prize squeeze and the pollution from the Chernobyl nuclear accident which is still affecting their farms to this day. The farmers have in fact taken charge of raising money and buying wind turbines that generate power for themselves as well as their local communities. The same applies for hill farmers who have restored old dam facilities to generate hydro power. At the same time, the phasing out sheep production has meant a revitalized ecology and the maintenance of wetlands that are an intricate part of the national parks system in the area.

The consensus of this session was that there is room for more diversity in counter-hegemonic activities. Chris Carlsson's (2008) book *Nowtopia: How Pirate Programmers, Outlaw Bicyclists, and Vacant-lot Garderners are Inventing the Future Today* is an inspiration that claims that every little step forward needs to be celebrated now rather than wait for a utopia in a hundred years from now.

After the session with the academic blockaders, we joined the big march of the day from the Højbro Plads to the Bella Center. It turned out to be the biggest demonstration in the history of

Copenhagen. Electronic music with a deep fast beat greeted us as we approached the demonstrators. It sounded more like a rave than a demonstration. We began to walk while looking at the marchers and the floats with speakers telling stories about the devastation brought on by climate change. We noticed a cobblestone on the ground. A woman on a bike asked us to give it to her. She told us that she would take care of it so it wouldn't be used for a bad purpose. At times, the march seemed festive, at other times somber as the riot police paced up and down the line of protesters. We tried to avoid them but somehow ran into them time and time again. Feeling uneasy we skipped out in Christianshavn and decided to observe the march from a distance. We climbed up the 400 steps of the 80 meter spire of Vor Frelsers Kirke, Our Saviour's Church, and got a good view. We watched from a distance as police arrived with vans flashing red lights to arrest over a hundred protesters who apparently constituted a security risk.

We were reminded of the drastic legal tools constructed by the Danish state in anticipation of demonstrations. Prior to COP15, a new "protest package" of laws had been passed which allowed the police to administratively hold people for 12 hours without a charge and detain people for a minimum of 40 days for disturbing the peace or disobeying the police.

Figure 5: Protesters at COP15 (photo by authors).

Figure 6: COP15 Demonstration on December 12, 2009, at Christmas Møllers Plads. In the distance, riot police arrest hundreds of protesters (photo by authors).

We also watched as riot police circled and patrolled Christiania, the now almost 40-year-old squatter settlement which promotes an alternative lifestyle of collective, car-free and sustainable living, in the middle of Copenhagen (Christiania, 2010). Spanning 34 hectares, Christiania is run on a self-governing basis, welcomes marginalized people, and contains many community-organized industries, businesses and shops. Many of the Klimaforum visitors were guests in Christiania. It represents perhaps the kind of vision and refuge that the world is calling for. We were drawn to Christiania on several occasions, simply walking the site, having a beer produced at the site, or taking in a vegan meal. When we asked for a beer at a restaurant, we were told they had none but that we could buy one at the neighbouring pub and bring it.

Figure 7: Christiania, Europe's oldest and largest squatter settlement, is Copenhagen's second most visited "tourist attraction" after the Little Mermaid (photo by authors).

13 / 12

We were reminded of the heavy police presence monitoring and control of COP15 when we followed up our association with the Academic Seminar Blockade the next day. We were part of a series of activist blockades at different sites in Copenhagen. Our assignment was to stake out a natural gas and oil-burning energy plant, the Svanemølleværket, owned by Dong Energy, in the north of the city. We gathered at Café Baresso at Triangeln in Østerbro and walked the 25 minutes to the site in Nordhavn. Already walking there, we spotted several police vehicles that were monitoring our movement. When we got there the place was buzzing with riot police. As we lined up in front of the gate, a concerned staff approached us on the inside, but we assured him that we were peaceful protesters who had no intention to scale the gate or destroy the property. Meanwhile, three vans of riot policemen parked in a parking lot some distance away from us watching our moves. We proceeded to present our papers, which ranged from theoretical and practical discussions about building action and solidarities and the role of the academe. It was difficult to concentrate on the speakers with the riot police around. As we spoke two of them visited us. One of them checked our nametags which we had put on randomly and therefore didn't tell who we really were. We finished the presentations with a collective chalk painting at the entrance which denounced the use of dirty energy and urged the teaching of climate justice. As we finished up, we saw the three vans of riot police leave.

Figure 8: Academic Seminar Blockage at Svanemølleværket, a natural gas and oil-burning energy plant in Nordhavn (photo by authors).

14/12
The next day, one of us attended one of an infinite number of sessions on the impacts of climate change on common folks in the developing world. The session, "The Impacts of Climate Change on People's Livelihoods in the Lesser Developed Counties", was hosted by LDCWatch (www.ldcwatch.org), a group formed at a United Nations conference in Brussels in 2001. It strives to unite civil society organizations in the North with counterparts in the South. On the panel were representatives from Haiti, Senegal, Bangladesh, Ethiopia, Korea, the Phillipines, East Timor and

host nation Denmark. Most of the presentations were in halting English, some were in other languages but were translated by what seemed like amateur volunteers (who were sometimes helped by members in the audience).

The moderator emphasized that climate change is not a technical or academic but a human rights issue. Action is needed immediately rather than in the future because people are dying now as a result of climate change. In Senegal, floods in Dakkar occur as a result of extreme rainfalls, yet extreme droughts are present too. Food production is thus unpredictable, which has led to greater imports and a growing national debt. The sea level is rising in places causing fishers to move. Fish processing activities, mainly done by women, have disappeared. The level of poverty is increasing and men are leaving for work elsewhere. The Bangladeshi representative told similar stories. Low-lying areas are flooded by the ocean and the number of environmental refugees is increasing, putting pressure on other areas and the cities. The farm community, which has fed a population of 150 million in the past, is now threatened. In Haiti, severe ecological disaster has arrived in the form of more severe and frequent hurricanes; rains were welcome in the past but are now feared. The gravity of the situation is worsened by the past process of deforestation associated with colonialism. Since COP15, the situation has been exacerbated yet again with the 2010 earthquake. In Ethiopia, chronic droughts have led to chronic food shortages and repeated bouts of starvation. In East Timor, water scarcities and new diseases have arrived while the national government is still pursuing a petro-based economy. The Nepali spokesperson recounted that the country's weather pattern has changed so that when it rains it is not needed and when it doesn't rain it is needed.

These Third World representatives called for climate justice now, compensation for past and current sufferings, and immediate action to help with the situation. Yet they also recognized the internal divisions within their home countries and acknowledged the conundrum of aid ending up in the wrong hands rather than to those in immediate need.

We attended another session led by the Durban Group on Climate Justice and its critique of carbon trading and how it fails both the developed and developing countries. Carbon trading is a complex and complicated issue. It is part of the Kyoto Agreement and supported not only by many big businesses and governments, but also some major international environmental organizations, such as Environmental Defence and Resources for Tomorrow — yet it fails miserably. The Durban Group demonstrated that the EU Emissions Trading Scheme, the world's largest carbon market, has consistently failed to cap emissions. They also showed that the United Nations' Clean Development Mechanism time and time again favours environmentally ineffective and socially unjust projects. Through the "Mechanism", many dirty industries in Europe and North America have been able to continue their operations by offsetting their emissions by investing in questionable operations in the Third World. Routinely, the latter operations displace local economic activities and livelihoods and are typically resisted and resented by local residents.

'Leave the oil in the soil' is a common rallying cry in the climate justice discourse. It corresponds with the view that the people who live in oil-producing regions seldom benefit from the resource but see it as a curse. The Durban Group added another set of measures, including a re-assessment of energy demands (which are always exaggerated), the expansion of regulatory, taxation and legal measures to deal with carbon emitters and offenders, a closer consideration of the concept of ecological debt, the securing of land tenures for Indigenous Peoples and forest-dependent communities, the promotion of local farming and food sovereignty, the support of local action, and the building of alliances between communities and movements based on local needs and desires (Gilbertson and Reyes, 2009: 92).

15 / 12

On the last day in Copenhagen, when we trekked back to the Bella Center to connect with our university colleagues to join in the launch of a Climate Justice website set up by climate change advocates at York University and the Ecoar Institute for Citizenship in São Paulo, Brazil (Climate Justice, 2009), we were again confronted with the growing restrictions to access and participation at the official Bella Center venue. We arrived a couple of hours before our arranged meeting time. However, this time most participants had arrived in Copenhagen and the pressure to enter the Center was at its peak. In fact, the organizers gave out a restricted number of second permits on December 14 to ease the mounting pressure on the Center. We waited in line for close to three hours without having the entrance door in sight. Tempers ran high in places but were eased a little by the police serving hot tea and coffee. The explanations for the melee that we heard in the line-up ranged from poor administrative skills of the organizers to sinister conspiracy theories that suggested a scheme plotting against environmentalist groups having sufficient insight at the venue. Whatever the cause, the effect was similar. Many participants felt that they weren't really part of the Bella Center deliberations any more.

But for us the line-up turned out to be an interesting experience in itself: it became an equalizer for many of the star participants and rank and file members of the environmental movement. We met and managed to get two exclusive interviews with Bill McGibbon of the 350.org campaign and Canadian Green Party Leader Elizabeth May who were also stuck in the line-up. McGibbon spoke to us about Canada being "as big as a carbon criminal as the world has" and May bemoaned the lack of progress at the negotiating table inside the Center (for a full account of the interviews, see Rabbletv, 2009). Both expressed a growing cynicism and pessimism about the Bella Center deliberations.

We said good bye to May and headed back to the comfort of the Klimaforum. We went straight to a session which featured George Monbiot (Monbiot, 2010), the *Guardian*'s columnist on environmental matters and author of the book *Heat: How to Stop the Planet from Burning* (Monbiot, 2007), who spoke eloquently about climate change. At current rates, carbon emissions threaten the survival of the earth and action is needed direly. Two fellow panelists from Europe stressed that peak oil is soon on us, the point where production will begin to decline. But Monbiot disagreed. There is likely more oil to be found and the world is flush with coal resources. Hoping for scarcities to prompt carbon mitigation is a pipedream; far too often dire predictions of resource depletion have been proved wrong. Monbiot thus argued that action is needed irrespective of present and pending scarcities.

Monbiot was particularly powerful in pointing to the contradiction in the behaviours of world political leaders on the question of climate change. At the same time as consumers are asked to cut carbon emissions and become more efficient in energy use, the very same leaders are supporting and subsidizing the continued search and production for carbon resources. Monbiot ended on a poignant note, endorsing the notion of carbon capture and storage as the most effective and efficient and cleanest and cheapest way to deal with carbon emissions. We gasped at this situation, knowing the impracticality and immense scale that carbon capture entails, never mind the spending of energy and effort to force it into the ground. However, Monbiot then told us what he really meant by the term: leaving the oil in the soil. How elegant.

The very last session at Klimaforum that we attended sealed the case for our assessment of COP15 as an utter failure. The Angry Mermaid Award was "set up to recognize the perverse role of corporate lobbyists, and highlight those business groups and companies that have made the greatest effort to sabotage the climate talks, and other climate measures, while promoting, often profitable, false solutions" (Angry Mermaid, 2009). The Award was sponsored by six non-government organizations, Attac Denmark, Corporate Europe Observatory, Focus on the Global South, Friends of the Earth International, Oilchange International and Spinwatch. There were 10 nominees: Monsanto and the Round Table on Responsible Soy; Royal Dutch Shell; American Petroleum Institute; American Coalition for Clean Coal Electricity; International Emissions Trading Association; European Chemical Industry Council; International Air Transport Association; and Sasol. The first three places went to Monsanto and the Roundtable on Responsible Soy (37.2 % of the vote), Royal Dutch Shell (17.9 % of the vote) and the American Petroleum Institute (13.8 % of the vote).

Monsanto received the award for lobbying for genetically modified RoundupReady soy to become a climate-friendly crop that qualifies for carbon credits. The company claims such soy can be grown without plowing which prevents the release of carbon. However, the monoculture crops require herbicides that damage human health and the environment, replaces valuable forests, and displaces rural and Indigenous communities.

Monsanto was a particularly worthy winner because of its efforts to lobby through the Roundtable on Responsible Soy, which consists of major environmental groups, including the WWF World Wide Fund for Nature. At the end of this part of of the award ceremony, an Argentinian environmentalist pointed to how some of the large environmental groups are working hand in glove with corporations.

Figure 9: The Angry Mermaid Award for corporate irresponsibility patterned after Copenhagen's most famous statue, The Little Mermaid (photo by authors).

Naomi Klein announced the award. She began her speech by telling us she was angry over the arrest of one of the main organizers of the next day's protests march to the Bella Center. She also told us that the hard questions about who is in charge and who has the power in the climate change debacle are not addressed at the Bella Center (for a full account of Klein's address, see Rabbletv, 2009).

As the closing event to our COP15 visit, the message of the mock awards was sobering. Light and entertaining as they were, they still presaged the crucial message that corporations are in charge of the climate change agenda. It's the corporations that are calling the shots in promoting and supporting the economic growth cycle and the flow of profits. They spend millions of dollars on lobbying efforts and they have the ears of the politicians.

The negotiation of nation states is more of less a shell game that fails to address the key issue: how to put a stop to the exploration of fossil fuels and put a break on their exploitation.

Figure 10: The International Emissions Trading Association (IETA) was one of the nominees for the Angry Mermaid Award. The Association claims to be an independent, non-profit, and dedicated to effective systems for trading in greenhouse gas emissions. The nominators felt a little differently about the Association's mandate (photo by authors).

Conclusion

Media portrayals of COP15 have been disparaging overall. The developed countries are invariably pictured as reluctant to reduce carbon emissions to the extent necessary to have a positive effect, nor prepared to pay compensation to those developing countries affected by climate change to no fault of their own. Conversely, developing countries such as China and India are accused of being too insensitive to tackle carbon emissions. The picture looks bleak. But the plight and solutions sought for in these portrayals are typically framed in the context of modernist institutions, which maintain the current power structures and material development aspirations in developed and developing countries alike. Solutions are thus sought in the sphere of the market and continued corporate and elite domination. Conversely, climate justice advocates tackle climate change at the level of who owns, controls, and profits from fossil fuel extraction, and who suffer the most from its consequences, both at the local scale of extraction and the global scale of atmospheric and climate impacts. Theoretical critiques and empirical studies exhibiting the bankrupt nature of modernist solutions as well as giving examples of transition societies show the way ahead. To do something about that situation is very difficult to be sure, but shying away from its presence is worse. We suggest that listening to the folks at the Klimaforum rather than the Bella Center is a step in the right direction.

* * *

REFERENCES

Angry Mermaid, 2009. www.angrymermaid.org, accessed 15 January 2010.

Carlsson, Chris, 2008. *Nowtopia: How Pirate Programmers, Outlaw Bicyclists, and Vacant-lot Gardeners are Inventing the Future Today*. San Francisco: AK Press.

Christiania, 2010. www.christiania.org, accessed 4 February 2010.

Climate Justice, 2009. www.globalclimatejustice.net, accessed 21 February 2010.

Gilbertson, Tamara and Oscar Reyes, 2009. "Carbon Trading: How it works and why it fails", *Critical Currents*, no. 7. Uppsala: Dag Hammarskjold Foundation.

Monbiot, George, 2007. *Heat: How to Stop the Planet From Burning*. Toronto: Anchor Books.

Monbiot, George, 2010. www.Monbiot.com, accessed 27 July 2010.

Rabbletv, 2009. www.Rabble.ca/rabbletv, accessed 8 January 2010.

NAOMI KLEIN

Paying Our Climate Debt

David Lewis Memorial Lecture.
Delivered by Naomi Klein, February 22, 2010, Toronto

Climate debt is a relatively new issue in the Global North, one that has received almost no coverage in the corporate media and only a first glance elsewhere.

Where it did enjoy a brief spotlight recently was in Copenhagen, at December's huge climate change summit, when it was championed by a coalition of Latin American and African nations, as well as a large number of NGOs from the Global South.

So, given the relative newness of the topic, it makes sense to start with a definition.

At its most basic, climate debt is the idea that poor countries are owed various forms of reparations from rich countries for the climate crisis.

It is also the idea that nature has rights, including the right to regenerate; that we have violated those rights and must now undertake a process of repairing the Earth.

The science underpinning climate debt is familiar to all of you here. The Earth's atmosphere has a finite carbon budget — a total amount of carbon that can be emitted by everyone before so much carbon accumulates that warming becomes catastrophic.

A safe level would be 350 parts per million; we have already passed that and we need to bring it down. In other words, we are over our collective carbon budget.

But the climate debt argument goes beyond that and points out that not everyone is equally responsible.

We in the highly industrialized north consumed far more than our share of the global budget. Our actions created a crisis, a crisis that is global in its reach and for which others are paying. For this we owe a debt on the basic principle of: "the polluter pays."

But the climate debt argument goes further still. The atmosphere is a common global resource on which no nation has any greater claim than any other. Using the atmosphere as a sink to absorb our emissions benefited our countries greatly: we got to build our profitable industrial bases and grow our economies.

By using up so much more than our share of the atmosphere we have robbed billions of people of their rightful share. That means they can't spew cheap dirty fuels like we did, a fact that has made development more expensive.

In short, we took what wasn't ours to take and for that we owe.

The end of Kumbaya

Climate debt differs from the ways of thinking about the environment that we have grown accustomed to because it is not, primarily, about polar bears. It is about people.

Traditionally, North American and European environmentalists have tended to sell climate change as a Kumbaya issue, one that erases difference: "We all share this fragile blue planet, so we all need to work together to save it. Everybody's got grandkids to worry about — even Dick Cheney."

In sharp contrast to this way of thinking the climate debt movement pointedly stresses *difference*. It zeroes in on the cruel contrast between those who caused the climate crisis (that would be us in the North) and those who are suffering its worst effects (that would be the Global South, most immediately Africa and the Island states).

Of course we will all be affected by climate change in one way or another but we are *not* all affected by it in the same way —

there are vast differences in the immediacy and intensity of the threat. For people in Yemen facing severe water shortages that fuel conflict, for instance, or for people in Bolivia watching their glaciers melt, climate change is not about grandchildren 60 years from now; it is about collective survival right now.

20% of the world's population is responsible for 75-80% of the historical emissions that created the climate crisis.

The dramatic spike in climate change denialism in the U.S. and Britain must be seen in this context: clearly many of us in the North do not feel the urgency of this issue, so we have all kinds of time to waste on denial.

Climate debt stats

Too much climate talk is bogged down in statistics and jargon and I really want to stay away from that tonight. But there are two key sets of numbers we do need to get our heads around.

The first set of numbers: 20% of the world's population is responsible for 75-80% of the historical emissions that created the climate crisis.[1] That 20% is comprised of the population of the industrialized, developed world — rich countries like Canada, the United States and Britain.

The reason we talk about "historical emissions" — the amount our countries have emitted since the industrial revolution — is because carbon stays trapped in the atmosphere, increasing in density, until it reaches dangerous levels.

So in determining who caused the climate crisis we can't just look at who emitted what last year or the year before — we have to take the long view.

Here is the next set of numbers: according to Justin Lin, chief economist at the World Bank, "About 75 to 80%" of the *damages* caused by global warming "will be suffered by developing countries" — the majority of the world's population.[2]

When you look at these stats side by side, you see that when it comes to climate change, there is a cruel inverse relationship between cause and effect — an utter disconnect between who has caused the crisis and who is living and will live its worst effects.

Breaking agreements we have already made

The idea that rich countries bear a much greater responsibility for the climate crisis is a principle our governments have already agreed to.

The United Nations Framework Convention on Climate Change, ratified by 192 countries including the U.S., recognized the concept of "historical responsibility" for climate change. That's why the Kyoto Protocol put the onus on developed nations to cut their emissions, at least in the first phase of its implementation.

The other reason why the Convention and Kyoto put the onus on developed countries is that the agreements recognize the connection between emissions and development. Countries like India, where 400 million people do not have basic electricity, will need to emit some carbon in order to achieve basic development goals.

About 75 to 80% of the *damages* caused by global warming will be suffered by developing countries — the majority of the world's population.

This is based on a recognition that not all emissions are morally equal: the emissions produced to electrify a rural village for the first time, for instance, are not the same as the emissions produced by a third car, or by a country that can't be bothered to invest in public transit, despite having ample resources.

As the prominent Indian environmentalist Sunita Narain puts it, there are "luxury emissions" and "survival emissions" — and they do not carry equal moral weight. Indians have a right to electrification, but we do not have a right to refuse to modify our consumption in any way.

Another way of thinking about it is the way Bolivian President Evo Morales puts it: the Earth does not have enough for the North to continuously live better and better, but it does have enough for us all to live well.

The privatization link

When the Climate Convention was written, it was not yet clear that the effects of climate change would fall disproportionately on the poorest nations, but it is clear now, making the case for climate debt even stronger.

Much of the reason why climate change is so much more of a threat to some is purely geographic: a 2°C temperature rise translates into an unbearable 3-3.5° increase in Africa. In Copenhagen, Archbishop Desmond Tutu protested that aiming for a 2° increase "is to condemn Africa to incineration and no modern development." And for low-lying island states like Tuvalu and the Maldives, geography is everything: as waters rise, their countries are literally disappearing.

But in addition to these flukes of geography, there is also the fact that the poorer the region, the less public infrastructure it tends to have to cope with climate change — whether it's the money to build strong seawalls to keep back rising tides, or the first responders to rescue the people when disaster strikes.

And it's important to recall that we in the North share much responsibility for this as well. That's because in so many cases, public infrastructure isn't weak by happenstance — it has been deliberately and systematically weakened by decades of what used to be called "structural adjustment": the policies attached to loans, our so-called aid, that required countries like Haiti and Rwanda to destroy their internal markets and privatize their public sphere in order to get a desperately needed loan.

Follow the debt: Haiti and Bangladesh

Haiti for almost its entire existence, has been carrying one form of odious debt or another — debts that were illegitimate because of their origins. First came the debt imposed by the French after liberation: claiming that Haiti owed France reparations for the loss of their former profitable slave colony, they threatened to re-enslave the country unless the new government agreed to pay a sum that would be worth $21-billion today. It took Haitians 122 years to pay it off.

Next was the debt racked up by the Duvalier dictatorships. When Baby Doc was forced out in 1986, he left behind debts of almost a billion dollars. These were odious debts because we know the money wasn't spent on Haitians, it went to the Duvalier's elaborate network of Swiss bank accounts and lavish properties on the French Riviera, as subsequent court investigations have found.

The point is this: for more than two decades, the country's Western creditors insisted that Haitians pay millions every year servicing those illegitimate debts — money that could have gone to build schools, hospitals, electricity lines, roads. The kind of infrastructure that makes countries more resilient when faced with disasters.

Worse than that, however, were the conditions applied to the loans, and are still applied to this day:

* Haitians were forced to drop their protections for local rice producers and open themselves completely to cheap foreign imports, decimating the countryside.
* Lenders demanded that Haiti privatize its telephone system and electricity.

Failure to comply was met with punishing aid embargoes, which dealt the final blow to Haiti's public sphere including its first responders: firefighters, police.

The effects of all of these policies are seen every time disaster strikes:

* the overcrowding in the cities because of mass migration from the countryside;
* the shoddy construction because of total deregulation (Haiti doesn't even have a building code);
* the absence of first responders; and
* deforestation, because when the state can't hook you up to electricity, charcoal fuel will have to do.

All this means that as the weather becomes more volatile, Haiti has little resiliency.

We saw it in 2008, when four strong storms hit the western Caribbean in 30 days. In Cuba they took the lives of four people. In Haiti they killed 800.

Bangladesh is another dramatic case. There, World Bank loans have been used to push farmers away from rice and towards export-driven shrimp farming on a massive scale. One side effect is that mangrove networks have been decimated, and mangroves traditionally form a natural barrier to cyclones in Bangladesh.

Now when increasingly fierce cyclones hit (linked to climate change) there is nothing to protect against massive land erosion. Large parts of Bangladesh are disappearing, creating an internal refugee crisis in an already desperately overcrowded nation. So this is about much more than geographical bad luck.

Fortressed climate green zones

The inequalities of climate change really clicked for me when I stumbled across something called the Climate Change Vulnerability Index. It is produced by Maplecroft, a private research outfit that provides high-priced information to corporations about which parts of the world are most vulnerable to global warming and therefore the riskiest places to invest.

They generate these color-coded maps of the world and I want to show one to you, since it had such an impact on me:

[Map not included in this article for the reason Naomi Klein explains.]

Forgive the bad quality — you have to pay thousands for the high res version. According to the key, the idea is that the lighter color a country is, the better off it is when facing climate change, and darker is worse. (We'll leave the in-depth analysis of the racial subtext for another time.)

There's Africa, very dark blue, not looking good.

There is the U.S., pretty light green, not bad.

But who is the biggest lightest green blob of all? Canada...

There are a few darker spots on the Canada map indicating a likelihood of droughts in southern Saskatchewan, fires and storms on the west coast.

But overall we fall into the absolute "lowest risk" category. The truth is there will even be some benefits for us, like much longer growing seasons. And valuable shipping routes opened up in the Arctic, thanks to melting ice.

Is it any wonder that climate change denialism is gaining ground in happy green countries like the U.S., Britain, and Canada? We can afford to waste this time, or at least we think we can.

And you can begin to understand why, from the perspective of the most affected countries, climate change looks less like a collective challenge that we are all facing together, and more like a silent war being waged by the rich against the poor.

Canada is a major aggressor in this war. *We* are doing this to Africa, to the Maldives, to Bangladesh, to Haiti — with our careless over consumption, with our scramble for the dirtiest sources of fossil fuels on Earth (the Alberta tar sands), with our failure to build other economic engines to sustain ourselves, and with our government's role in actively sabotaging the possibility for a serious climate agreement in Copenhagen. We are not passive; we are aggressors. Our actions affect others.

I know environmentalism isn't supposed to sound like this. It's supposed to be green and fuzzy, not red and angry.

But climate change is not just an environmental issue anymore. It is the human rights issue and social justice issue of our time.

And it's time to talk about paying our debts.

What are climate debt reparations?

In Copenhagen, I was part of a side conference on climate debt, speaking on a panel about reparations. It was an attempt to get specific about what climate reparations should look like.

I was struck that almost none of what was proposed was about money.

For instance, one of the panelists was a minister from Bangladesh. He was calling for a total overhaul of international refugee law, arguing the West needed to open its borders to climate refugees — not out of the goodness of our hearts but because it is the least that we owe to people who are losing their lands because of our actions.

But climate change is not just an environmental issue anymore. It is the human rights issue and social justice issue of our time.

That said, climate debt is also about economic resources and real transfers of wealth. Adapting to a changing climate carries heavy costs: developing countries need money to build seawalls and better disaster response infrastructure. They also need money and technology to leapfrog over fossil fuels and go straight to cleaner alternatives.

Angelica Navarro, Bolivia's chief climate negotiator, who was also on the panel, has called for "a Marshall Plan for the Earth." As she puts it: "This plan must mobilize financing and technology transfer on scales never seen before. It must get technology onto the ground in every country to ensure we reduce emissions while raising people's quality of life. We have only a decade."

What does that mean in practice? Well, for starters, those resources would mean that when Indian rural villages are electrified — which is their right — it isn't through coal-fired plants, but through a national solar network, decentralized and controlled by communities.

Another example of what kind of possibilities would open up if we recognized the existence of these debts is the battle over Ecuador's Yasuní National Park. This extraordinary swath of Amazonian rainforest, which is home to several Indigenous tribes and a surreal number of rare and exotic animals, contains nearly as many species of trees in 2.5 acres as exist in all of North America. The catch is that underneath that riot of life sits an estimated 850 million barrels of crude oil, worth about $7 billion. Burning that oil — and logging the rainforest to get it — would add another 547 million tons of carbon dioxide to the atmosphere.

Two years ago, Ecuador's center-left president, Rafael Correa, said something very rare for the leader of an oil-exporting nation: he wanted to leave the oil in the ground. But, he argued, wealthy countries should pay Ecuador — where half the population lives in poverty — not to release that carbon into the atmosphere, as "compensation for the damages caused by the out-of-proportion amount of historical and current emissions of greenhouse gases." He didn't ask for the entire amount; just half. And he committed to spending much of the money to move Ecuador to alternative energy sources like solar and geothermal.

For a while the project seemed to be getting off the ground but Correa has recently called the whole thing into question, saying foreign donors were trying to exert too much control. And now he is threatening to drill.

These are just a couple of the kinds of things climate reparations could pay for. There are thousands of others. But we should be clear: the sums needed are big. There is no agreement about how big but a team of United Nations researchers puts it at $600 billion a year over the next decade.

These are scary numbers but keep in mind that this is not a zero-sum game, with money disappearing into the ether with no benefit to us. We all benefit from avoiding catastrophic climate change, from saving the Amazonian rainforest. After all, it's no fun living in a green zone, surrounded by a sea of raging red.

Paying our climate debts, and being part of this global process of repairing the Earth — making it more balanced economically and ecologically — is not dour punishment. It is, in fact, the most inspiring collective work we could ever hope to undertake.

The R-word

In Copenhagen and afterwards, quite a few U.S. environmentalists have approached me to ask me — straight up — to please stop talking about debt and reparations. Here's a typical reaction to climate debt from the Executive Director of International Rivers, writing in *Huffington Post*: "Sarah Palin, Glenn Beck and the rest of the American Taliban would surely wet themselves with delight if there was any serious noise from the left calling for billions of dollars in climate reparations for poor countries."

Being part of this global process of repairing the Earth — making it more balanced economically and ecologically — is not dour punishment.

And there is no doubt that climate reparations are a tough sell in the U.S. So I want to talk a little about why these words, however controversial, are important.

The first reason is because they are true.

We have done great damage — to the earth, to our fellow human beings — and we must begin the process of healing and repair. Every reparations process in history has begun with an acknowledgement of past wrongs.

There are of course practical reasons too. One way or another, billions of dollars will be spent in the name of helping countries cope with climate change.

If it's *not* recognized as reparations, then it's simply aid, and we know the way aid works. Much of it will come in the form of loans, not grants.

Much of it will be phantom aid, obtained by robbing money from school construction and AIDS programs to pay for climate mitigation. And it will be administered by the usual suspects, particularly the World Bank, already gunning for this powerful role.

If it is aid, as opposed to reparations, then we, the donors, remain all powerful, able to dictate whatever terms we like. In

The Shock Doctrine I discuss how aid in the wake of natural disasters is frequently used this way.

And it is already happening in the climate negotiations: the scandalously weak "Copenhagen Accord" — pushed through at the last minute by the U.S. and a handful of other states — is being used to bribe developing countries to drop the entire concept of climate debt.

Apparently $100 billion will be on the table for developing countries by the end of the decade to mitigate the effects of climate change. But countries are being told that they won't get any of it unless they endorse the weak accord.

It is naked blackmail — forcing developing countries to choose between a strong fair deal that stands a chance of averting climate chaos and the funds they need to cope with the droughts and floods that have already arrived.

And this is the most important reason why the climate debt frame is so important and cannot be sacrificed in the name of political pragmatism. If we are paying our debts, paying for a crisis we caused, then our governments come to the table with some humility, without the insistence on dictating the terms.

The countries receiving the resources, on the other hand, come to the table not as perpetual deadbeat debtors or for a handout, but as creditors, with their own claims to make.

It changes the very concept of Who Owes Whom in the global economy.

Debt is about power

By now more than a few of you are probably thinking: this is all very well and good but it's not the way the world works.

And over the past few months of raising the issue of climate debt in various forums, I have found a consistent unwillingness to argue the facts, but rather a tendency to dismiss the whole idea as "untethered to reality" as Obama's chief climate negotiator Todd Stern put it.

Basically the message is: you may have the facts on your side, but you lack the power. And debt is always, we must recall, about power — it's about who has the muscle to get their debts collected. That muscle can come in the form of broken kneecaps, if you are the mob, or the ability to break a country, if you are the International Monetary Fund.

Those who dismiss the climate debt argument are essentially saying:

"You and what army? How are you going to make us make good?"

If you have no power, you get no payback.

So if there is such a thing as a climate justice movement, and I believe there is, and it is growing, its job is to somehow change these poisonous power dynamics — inside our countries and between nations. And the way we change these dynamics is not with raw force, as our adversaries do. It is through organizing: by putting some movement muscle behind these legitimate demands.

I like the way Bill McKibben, a wonderful writer and founder of 350.org, puts it.

> Our goal... is to get all the support we can behind the most vulnerable poor nations on the planet. They're showing unbelievable courage in sticking up against the pressure from the U.S. and others... They're demanding their survival be taken seriously. We're trying to provide the armies that they lack... mobilizing civil society behind the science and behind survival. So that's what... the next couple years will be about.

Tar sands: bringing climate debt home

The climate debt movement in the Global South has been very clear that the most important form of reparations is not money. It is stopping the crime in progress by dramatically cutting our emissions. I'll quote Ivonne Yanez, an Ecuadorian activist with Accion Ecologica. In Copenhagen she told me: "It's not only the question of the polluters pay, but the polluters need to stop what they are doing. That is justice."

In Canada that means one thing: shutting down the crime scene known as the Alberta tar sands.

Then beginning the process of repairing that unspeakably scarred part of our country. And making reparations to the First Nations communities living down river who are facing alarmingly high cancer rates and who have faced federal and provincial governments unwilling to listen.

If we are to have any hope of achieving those ambitious goals, we need to focus on why our government has felt free to thumb its nose at the world when it comes to Kyoto.

But first the stats:

Under Kyoto, Canada pledged to cut emissions by 6% between 1990 and 2012. Rather than cutting we have increased our emissions by 35% since 1990. The Harper government appears to be daring the rest of the world to stop them. No other country has a record this abysmal. There is one reason: the tar sands.

This kind of defiance is not the way serious international agreements usually work.

Normally when governments break international agreements they hear about it. If we in Canada subsidize our logging industry, the U.S. hauls us to trade court, charging that we violated NAFTA. If we ban a harmful additive to gasoline, same thing: the U.S. hauls us to trade court, and threatens us with a trade war.

Canada pledged to cut emissions by 6% between 1990 and 2012. Rather than cutting we have increased our emissions by 35% since 1990. No other country has a record this abysmal.

But when it comes to the environment, there is clearly a lack of political will to enforce the agreements we have signed. And when governments make a mockery of international law like that, then there is only one thing that can fill the gap: social movements, civil society, launching boycott and divestment campaigns that exact real costs from the offenders.

It is already starting to happen.

- Whole Foods and Bed Bath and Beyond have both announced that they will boycott fuel coming out of the Canadian tar sands, which required five times more energy to produce than traditional crude.
- California has adopted a low-carbon fuel law and other states may follow.
- The Pentagon is starting to cut back on its consumption of tar sands oil to meet new requirements that government agencies reduce their emissions.

I mean when Bed Bath and Beyond and the Pentagon are lined up against you — that's bombs and bath tubs — even Alberta has got to start paying attention.

Most significantly, the oil corporations themselves are being targeted directly: Shell and BP are both facing anti-tar sands shareholder resolutions at their upcoming AGMs.

And much of the anti-Olympics activism that has been taking place in Vancouver has focused on outing the dirty secrets of two of the games' main sponsors: Royal Bank and Petro Canada. Both are up to their necks in the tar sands.

It brings back the days of the anti-Apartheid struggle, doesn't it? I always remember when my brother Seth, age 15, marched to Royal Bank and closed his account.

Perhaps it's time to cancel our RBC accounts once again, this time to prevent the future of climate apartheid that we appear to be hurtling towards.

Conclusion

Here in Canada, our government has turned our nation into the world's leading climate criminal for two reasons: because it is profitable — and because there are no consequences for its lawlessness.

So our task should be clear: we have to impose consequences from below, by making investing in the tar sands a liability for Big Oil, not a windfall.

Of course the consequences for the corporate world won't end there, and they reach far beyond our borders.

When we talk climate reparations, what gets people's back up is the idea that they are personally going to pay. But why should that be? Climate reparations don't have to come out of the pockets of regular taxpayers.

The oil and gas sector, big coal, agribusiness, and the banks that finance them — the players actually responsible for both climate change and underwriting the climate change denial movement — can be made to foot the bill.

With the right kind of taxes and penalties, the oil and gas sector, big coal, agribusiness, and the banks that finance them — the players actually responsible for both climate change and underwriting the climate change denial movement — can be made to foot the bill.

Remember: Exxon Mobil made more profits every year over the last three years than any company in history. I think they should pay their fair share of the global clean up.

Climate debt is not punishment; it is our best chance of securing the resources for repair and transformation that our planet needs desperately. Sometimes it feels awfully good to come clean and settle up.

* * *

Naomi Klein *is an award-winning journalist, syndicated columnist and author of the* New York Times *and #1 international bestseller,* The Shock Doctrine: The Rise of Disaster Capitalism *(2007). Published worldwide in September 2007, The Shock Doctrine is being translated in over 25 languages. Her first book* No Logo: Taking Aim at the Brand Bullies *(2000) was also an international bestseller, translated into over 28 languages with more than a million copies in print (from Naomi Klein's website: www.naomiklein.org).*

Copyright Naomi Klein 2010

ENDNOTES

[1] Redman, Janet, "Blame Game Leads to Climate Deadlock in Bonn," http://www.ips-dc.org/articles/blame_game_leads_to_climate_deadlock_in_bonn; Bolivia, "Climate debt: The basis of a fair and effective solution to climate change," Presentation to Technical Briefing on Historical Responsibility in Bonn, PDF, slide 5.

[2] "Climate Change Hits Poor Countries Hardest: WB," *Agence France Presse*, 4 October 2009. http://news.yahoo.com/s/afp/20091004/sc_afp/worldbankwarmingclimateeconomy.

Vandana Shiva Talks About Climate Change

AN INTERVIEW BY TOR SANDBERG

During the G8/20 conference in Toronto in June 2010, when the world leaders of the largest carbon emitters in the world assembled to discuss the global financial crisis, the Council of Canadians organized an alternative conference entitled "Shout Out for Global Justice." At the event, world-renowned Indian scholar and activist Vandana Shiva spoke. I caught up with Dr. Shiva during the day to ask her about her position and thoughts on climate change. The following is a slightly edited transcript of the interview. The on-camera interview is posted on rabble.ca under the heading "Dr. Vandana Shiva: G8/20 created to silence global majority."

Tor Sandberg: Climate debt — it's a term that's entered the discourse of environmental thought. Could you comment on that term and about the movement behind it?

Vandana Shiva: Well in my perspective, the minute we start pulling out fossil fuels from underground and we start to burn millennia of photosynthesis in a day, we are creating a climate debt to nature. Because the fundamental problem with climate change is we are abusing nature's cycle of carbon absorption and renewal. And then there is the second kind of climate debt, and that is related to the fact that the countries that are rich today, industrialised on the basis of access to coal, ... they externalised the impacts of this pollution to the whole world. This debt has not been paid and even today there is a reluctance to pay it. The entire collapse of the discussions at Copenhagen and the substitution of a legally binding treaty with a so-called Copenhagen Accord is one more example of one outstanding climate debt that hasn't been paid.

TS: Some within the mainstream environmental movement have proposed carbon trading as a system that could help curb and stymie climate change. What are your thoughts on this?

VS: When the very idea of emissions trading and carbon trading started, I thought it was as perverse as when the Catholic Church sold out indulgences where, if I was a sinner, I could keep paying the Bishops and continue to sin. In a similar way the polluters just pay off and continue to pollute. Carbon trading is to today's economy what indulgences were to the Catholic Church, and they're equally inappropriate. But we now have the record of how they've performed, and with the experience that we've had since Kyoto was signed, and since the emissions trading schemes of Europe and the U.S. were put in place, and since the Clean Development Mechanism between the North and the South was put in place, the two things that have come through totally (my book *Soil No Oil* revealed the fact that most Clean Development Mechanism funding has gone to dirty industry, ..., it has gone to CFC plants, it has gone to the steel sector, it has gone to mega dams. This is not clean development by any imagination. The second is that instead of reducing emissions while these carbon trading schemes were carrying on, emissions have increased by 16%, and the polluters have walked away with billions and billions and billions [of dollars]. I know that, for example, Mittal, who now controls most of the steel sector of the world, walked away with 1 billion pounds a year of pollution rights because governments are according large pollution rights to the biggest polluters, so, in effect, the atmosphere, which is our commons, has been divided up as property of the world's polluters and that's just not the way to go about cleaning up a polluted part of the environment.

TS: If you 'leave the oil in the soil', how will nations develop and their economies thrive?

VS: I think the first thing is to assume that pollution is development is just a wrong correlation. I come from India which until 10 years ago was what I would call a biodiversity economy, very tiny amounts of fossil fuel use. We were fine. We had food. We had water. As this dirty development which is called development has accelerated in India, we have lost our rivers, we are losing our forests, half of India is starving. This is not what development is supposed to be. I think this narrow-minded equation of "development is equal to fossil fuel use" should be given up for three basic reasons. First, in any case, this fossil fuel will not be around for too long, so if you want to have development for the next century, you'd better plan without fossil fuels because you have to push to riskier levels of exploitation and we've reached peak oil. We wouldn't have the Deep Water Horizon crisis if we had realized that it's time to phase out the fossil fuels. Canada wouldn't be mining its tar sands if we'd realized, ok, you know we've mined the … fossil fuels where they were easily accessible. You don't have to ruin an entire ecosystem. The second reason why this correlation should be given up is because, anyway, externalities are costing us a lot. Just last year, the Indian economy caused an extended drought that led to rainfall failure in the period when we should have had the monsoon. And when the monsoon should have passed we got such heavy rain in areas that are supposed to be dry areas that there was flooding — three hundred people died, the overall bill of the damage was 30 billion dollars. That's one year, one country. If this externality was being included, we couldn't afford a fossil-fuel driven path. And the third reason why it's totally inappropriate is because, you know, I really believe we have one option is an economy of fossil fuels, an economy of oil, and the other option is an economy of soil and an economy of biodiversity. And I think we are much better off in an economy of biodiversity. You have happier people. You have more basic needs met. You have more peace. In every indicator of what is authentic development, a non-fossil fuel-based economy is better off. I advise the government of Bhutan and they have evolved a different criterion of how they'll measure their progress. They don't measure gross national product, they measure gross national happiness. And on those real indicators, which tell you how people are, not how well corporations are doing, we'll be fine without fossil fuels.

TS: You're here now during the G8/20. What do you think of these kinds of meetings? How do you place their existence in the fight against climate change and better development?

VS: Well, the G7 was created when the South started to get unified to create a new economic order, which would be different than the North-dominated industrialised country model of the world. That's when the G7 countries said "we won't let the South create an alternative". Then we had a total collapse of legitimacy of the G7: everyone said seven men getting together is not the world economic summit. So what they did was bring in the countries that they have now, in a way, sucked into the global economy as carriers of the burden, particularly the environment burden; I've called it the outsourcing of pollution. So when Canadian companies, U.S. companies, are now present in India, if they have to make sure the decisions they want for all their corporations are implemented by the government of India, they have to suck in the Chinas and the Indias. We saw how this sucking in led to the erosion of the climate summit. The Copenhagen Accord was pushed by Obama, entering a green room like this, where the Chinese were sitting, and imposing on the four countries an accord which these five signatories took to the rest of the plenary and the rest of the plenary said: "but we didn't agree to it." I think countries like India are being abused and used. Unfortunately, our leaders feel so happy being at the rich man's table and they forget that this is purely facilitating the exploitation of our resources and exploitation of our people, and the destruction of our democracies.

TS: Could you speak about the Canadian government's role during the G8/20 meetings?

VS: You know the longest I have spent out of India ever was when I was a graduate student in Canada doing my PhD over three years at Western. And my experience of Canada, my experience of Toronto, my experience of London has been of a peaceful place. To see Canada militarize itself and militarize the city at the level at which it has, can only be a result of a deep separation of a government from its people, and a government from its own democratic culture. The blowing up of nearly $2 billion, just for security, to terrorize a city — I think is so immoral and unethical in a period of financial crunch and economic slow-down. If you've got to have cutbacks, it should be cutbacks on expenditures on summits like this, it should definitely be cutbacks on security.

* * *

Tor Sandberg *is one of the editors of this book.*

Dr. Vandana Shiva *is a leading independent thinker and voice for the South. Her work highlights the connection between human rights and protection of the environment. She is author of many books and publications, including* Monocultures of the Mind: Biodiversity, Biotechnology and Agriculture *(1993) and* Soil Not Oil *(2008).*

SONJA KILLORAN-MCKIBBIN

The Path from Cochabamba

In order for there to be balance with nature, there must first be equity among human beings.

— *Cochabamba Declaration*

From April 19-22, 2010, tens of thousands of people from around the world gathered in Cochabamba, Bolivia for the World People's Conference on Climate Change and the Rights of Mother Earth. Despite these record numbers, the international press was largely silent in stark contrast to the Copenhagen Summit (COP15) held in December, 2009, which dominated the news for days. Unlike COP15, this conference on climate change billed itself as a people's summit, with representatives from social movements, unions, and peasant organizations alongside academics and government representatives. Rather than involving business-style negotiations, the Cochabamba conference was organized for those most affected by climate change to respond to the threats they were and are facing. It was a forum for a rarely-heard narrative of climate change: not one of technological and market fixes but one that demands radical solutions. The 17 working groups at the conference defined different priorities from those that have dominated the discourse of climate change thus far and insisted upon a radical response, grounded in a new conception of development.

Figure 1: Crowds gather at the stadium in Tiquipaya, Cochabamba for the opening of the summit (photo by author).

The People's Conference diverged from previous climate change talks in three significant ways. First, there were groundbreaking levels of participation both in terms of attendance, and in meaningful venues for discussion and debate. Second, it addressed the structural causes of climate change and rejected all solutions that were rooted in the capitalist system. Third, the seriousness seen at Cochabamba was a refreshing change from previous discussions around environmental degradation. This seriousness was evident both in the analysis of the participants who are currently living the effects of climate change and in the demands resulting from the summit. However, many challenges lie ahead in implementing the vision from Cochabamba.

Whose summit? Whose climate?
The faces of climate change

Any story is constructed by those who tell it; the narrators not only choose what to emphasize or elaborate but where the story ends and begins. In the stories of climate change, we see that

Copenhagen was content to begin with a planet in crisis, facing disastrous levels of CO^2 emissions and looking for solutions. Defining the story in this way limits the discussion to the reality of the present system, almost inevitably leading to market-based solutions. Typically, the story of climate change speaks in abstract terms about the impacts and nebulous consequences that shall be felt in some far-away future.

In contrast, the Cochabamba Conference was not restricted to scientists, academics and policy-makers, but open to all. The initial call for participation in January stated that: "those most affected by climate change will be the poorest in the world who will see their homes and their sources of survival destroyed" (Morales Ayma, 2010) but expressed confidence that these same individuals could work in solidarity to come up with solutions. In this regard, the conference was undoubtedly a success. More than 35,000 people (approximately 10,000 from overseas) from a wide-range of backgrounds and classes registered for the conference and participated in the events, a level of participation previously unheard of in an international conference on climate change.

By including speakers who were feeling the impacts of climate change in their day to day lives and facing the stark reality of losing their water sources or being forced to migrate, the conference told the reality of climate change. This broad participation, which allowed for a multiplicity of narrators, demanded a much earlier beginning to the story of climate change. The story at Cochabamba began with the advent of carbon emissions, with industrialization, modernization, and growing inequality: in short, the history of climate change is the history of capitalism.

Moreover, delegates were not passive participants in the events. The conference declaration was the product of three days of debates and negotiations that were open to all participants. The 17 working groups were each responsible for creating a document that would outline the need for change in their respective areas: structural causes, harmony with nature, rights of mother earth, organizing a climate referendum, a climate justice tribunal, climate migrants, Indigenous rights, climate debt, shared vision, the Kyoto protocol, adaptation, financing, development and technological transfer, forests, the dangers of the carbon market, action strategies, and agriculture and food sovereignty.

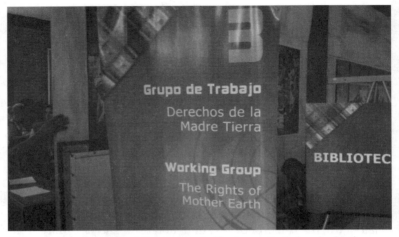

Figure 2: Crowds gather at the stadium in Tiquipaya, Cochabamba for the opening of the summit (photo by author).

These themes allude to the radically different vision from the outset of the Cochabamba summit. COP15 focussed on establishing a carbon market, financing adaptive measures, and establishing global emissions targets. These three simple objectives sought to regulate climate change but did nothing to address its origins. Cochabamba, on the other hand, did not limit itself to the realm of regulation but demanded a proposal that would recognize the gravity of the situation, challenge the root causes of climate change, and respond to the needs of the majority of the world's peoples.

This difference in scope reflects who was at the table. More important than the number of people at the conference was the dynamic level of participation in the working groups and the space for voices from traditionally marginalized groups. A representative from the Inter-Tribal Council of Alaska spoke of the devastation experienced by his peoples and the resettlements of Indigenous

Figure 3: Woman speaks at a working group on her experience of climate change (photo by Julien Lalonde).

80

populations already underway due to climate change. Participants from Bolivia spoke of their receding glaciers, changing weather patterns, and threats to life and livelihoods caused by diminished access to water. One of the most striking moments was hearing an elderly Indigenous woman speak about the impact of climate change on her crops. She became angrier and angrier as she spoke, insisting that her way of life was threatened because of the over-consumption of others.

This participation was a key element of the conference and spoke to the urgency of the situation. The message was clear: not only are the immediately destructive effects of climate change being felt in the present day, they signal what is to come to other areas of the world. The debates and discussions at the working groups were often heated, emotional and exhausting. That said, for once there was a space to hear all of the voices in the room. Participants weren't speaking just in abstract or theoretical terms, but of their lived realities and the direct impacts they were seeing everyday in their communities. With this level of participation, it is no wonder that Cochabamba represented a radically different narrative of climate change than those which came out of Copenhagen.

Figure 4: Thousands of people attended the closing ceremonies in Cochabamba (photo by Raul Burbano).

Figure 5: The Working Group on the Structural Causes of Climate Change (photo by author).

System change, not climate change: the structural causes

This new narrative situates the origins of climate change much earlier than typically thought. Rather than being a phenomenon of the past 20 years and linked to increased carbon emissions, climate change is seen as a symptom of a deteriorating way of life and the destruction of the relationship between human beings and the natural world as a result of capitalism. That climate change is tied to the larger economic system in which we live is painfully obvious to some, but is completely absent from discussions of carbon trading and offsets. This concern was central to the declaration that came out of Cochabamba which states that we are entering these moments of crisis because we have reached the limits of the planet due to the interminable growth of capitalism.

At the conference, Professor of Plant and Soil Science and Director of the *Monthly Review*, Fred Magdoff, spoke of capitalism's unending quest for growth, and the resulting need to commodify nature. Given this tendency, the only mechanism to protect the environment is to establish a system that no longer normalizes unending competition but promotes equitable distribution as the key to reducing over-consumption. Vice-President of Bolivia, Alvaro Garcia Linera, built upon this framework to insist that there could be no isolated or individualized solutions but that confronting climate change depended on a "revolution in all places" because climate change was inherently linked to the issue of power: who has it and who does not.

It is well known that the effects of climate change are most felt by the populations that have historically done the least to cause it. This also means that those affected have the least capacity to prevent it. The call to Cochabamba cited that 75% of historical emissions of greenhouse gases have come from the Global North in its pursuit of a "path of irrational industrialization" (Morales Ayma, 2010). The conference challenged these issues of power by allowing broad participation from wide-ranging groups but questions remain about how to truly confront these imbalances.

The logic of production and ever-increasing consumption has worsened inequalities and slowly turned all goods, from human labour, to land, to water, into commodities for trade and profit. The overexploitation of these common goods has come at a tremendous cost to citizens of the Global South. Capitalist expansion has brought about the increasing commodification of all goods — nature is no longer valued for its cultural, ecological, emotional value but instead for its contribution to production. The goods produced are completely divorced from their social, ecological and cultural contexts — we think not about the conditions under which the cotton was grown for our t-shirts or the working conditions of those who produced it; instead we understand products merely in relation to money. This process is constantly being extended to objects previously considered part of nature as can be seen with the patenting of traditional crops or the commercialization of water (Kranjc, 2006).

Sadly, many environmental proposals focus on a form of green developmentalism which seeks environmental protection through commodification (McAfee, 1999). Even climate change has been converted into a market. For example, carbon trading schemes represent the commodification of our atmosphere. We can see further environmental commodification in green products, the promotion of so-called ethical consumption, and clean-development mechanisms. The Cochabamba Declaration labels all these as false hopes in combating climate change. The same processes that have converted nature into natural resources cannot save us from the ongoing consequences. Instead, movements must confront commodification and strive for extra-market solutions that are based upon respect, reparations, and redistribution.

The growing people's movement identifies that environmental change can only be addressed through attacking structural inequalities by coming together around the proposal of *vivir bien* or "living well". This proposal is frequently contrasted with the dominant ideal of *living better* — the endless pursuit of more and more, whether it is more prestige, more wealth, or more goods. Living well is based on the Andean understanding of the lifecycle and connection to the earth around an inclusive model of growth and community links rather than individual capabilities (Prada, 2010). This slogan has been taken up by many groups around the world but there is still a great deal of uncertainty as to what it means, and how it can be implemented. In Bolivia much of the program consists of strengthening community production and confronting the supremacy of modernity. However, despite the popularity of this slogan, it is clear that confronting climate change and pervasive inequalities is about more than a shift in mindset.

Living Well	Living Better
• Strengthens communities	• Strengthens the individual
• Inclusive	• Exclusionary
• Redistribution of wealth	• Concentration of wealth
• Development planning from the community	• Planning from dominant hegemonic culture
• Self-determination, dignity and sovereignty	• Dependent development
• Reciprocity, bartering, complementarity	• Supremacy of the market
• Cyclical world-vision	• Linear world-vision
• Harmony with Nature	• Dominion of man over nature
• Goal: a fraternal social communitary life with solidarity	• Goal: individualism and monopolies with a widening gap between the rich and poor

Figure 6: Concepts of Living Well vs. Living Better from Bolivia's National Development Plan.

What do we mean when we talk about climate change? Seriousness of the proposal

A fundamental difference of this summit can be seen in the seriousness of the discussions and outcomes. Copenhagen proposed a series of negotiations to agree on measures that would address climate change, thereby demonstrating a flagrant disregard for its actual impacts and a presumption that the effects could be diminished through bargaining. In Cochabamaba the immediacy of the need was felt in every moment and was evident in the proposals presented. The conference was a space to analyze systemic causes of climate change and define strategies for action including four central objectives: 1) a Universal Declaration for the Rights of Mother Earth; 2) new commitments for future COP meetings around themes of climate debt, climate migrants, emissions reductions, finance, etc.; 3) a people's referendum on climate change; and 4) a Climate Justice Tribunal.

These goals are what Bolivia is now bringing to the table in subsequent climate change talks. The Universal Declaration for the Rights of Mother Earth parallels the Universal Declaration of Human Rights as a framework to recognize the earth as a living

being with inalienable rights. This declaration would provide an operating framework for the climate justice tribunal that would enforce environmental commitments and ensure accountability. Along with this mechanism for enforcement, the agreement seeks to keep global temperature increases below 1 degree Celsius. To

Figure 7: Evo Morales Speaks to Reporters on the Second Day of the Conference (photo by Raul Burbano).

achieve this goal, countries must honour the Kyoto Protocol as with an amendment for the period of 2013-2017 during which time emissions must be reduced by 50% of 1990 levels leading to reductions of 100% by 2040.

Discussions at Cochabamba examined the issue of environmental degradation from multiple levels. At the local level, there was widespread recognition among Bolivians that there remained much to be done. In his opening speech, President Morales sought a vindication of traditional goods as more environmentally friendly than their modern counterparts. It was clear, from his perspective, that a response to climate change not only depends on a structural attack on capitalism but increased accountability and a reconnection with the earth. Even though developing countries are among the lowest emitters, Morales encouraged Bolivians to look to their own life to revalue their connection to the land and reject patterns of consumption.

Figure 8: An educational display promoting the variety of native crops in Bolivia (photo by author).

But the People's Agreement does not limit itself to individualized responses to environmental concerns. Another key component was payment of the historical debt that the Global North owes the Global South. At COP15, aid was promised to developing countries to assist with costs of mitigation and adaptive measures. However, this commitment was not honoured as the United States quickly reneged on its promise of $100 billion by 2020 and then further cut funding to countries that had rejected the Copenhagen accord for not going far enough.

The People's Agreement moves away from this aid framework. Instead, it demands 6% of the GDP of developed countries as payment for the theft of environmental resources to date. This funding of 6% far surpasses previous commitments of international aid for climate change. More importantly, by considering these commitments as part of climate debt — or the equitable redistribution of atmospheric resources — the Cochabamba agreement changes the language of climate change responsibility and the framework for negotiation. This new framework of environmental governance covers all the bases from financial mechanisms, commitments on emissions reductions to the creation of an enforcement apparatus.

But there's always another story

Despite these strengths, it would be wrong to idealize what took place in Cochabamba. There were many problems with the conference, some apparently superficial — like the logistical difficulties that came from putting together a conference in just four months with little funds — while others spoke to more troubling divisions and contradictions.

Although it is undeniable that the level of participation was significant and wide-ranging, there were some notable absences. First, the National Council of Ayllus and Markas of Qullasuya (CONAMAQ), which was excluded from the conference because it challenged the contradictions between the Bolivian government's national environmental discourse and actions. They organized a separate working group, dubbed *Mesa 18*, to protest Bolivia's lax mining regulations and the ongoing environmental destruction taking place because of extractive industries in the country. Since the election of Evo Morales, Bolivia has adopted an ambitious development plan which seeks to reduce external dependency but continues the almost exclusive reliance on extractive industries to fund social programming.

Bolivia's situation reflects that of many other developing nations that have relied upon resource extraction for their development. The nationalization of key industries in Bolivia has improved national revenues from these sectors but the government has done nothing to move away from this dependence. At this point in time, the country appears to be carrying out a dual development program of industrialization and resource extraction on the one hand with the discourse of community development and valuing mother earth on the other. These contradictions call into question how serious the Bolivian government is about changing the development paradigm when it continues to be so reliant on extractive industries.

Likewise, partly because of this, there was an absence of labour movements at the summit. Although certain Canadian unions sent delegates, there was little representation of trade unions from the Global South. Bolivia had few delegates from the Bolivian Workers Central (COB) and other workers were poorly represented. Once again, this alludes to the growing gaps between the discourse of respecting mother earth and the government's political economic program. Pablo Stefanoni, editor of the *Monde Diplomatique* in Bolivia, condemned the conference for being representative of the government's "expanding hiatus between discourse and reality", provoking numerous responses from government and Indigenous groups. The ensuing debate, along with the general worker's strike that took place in early May, highlights the conflicts in the country.

Another notable absence was that of government officials from Northern countries. Although it was important to see such wide-ranging grassroots participation, one was left with uncertainty about the capacity of participants to implement the Cochabamba agreement; that is to move it beyond a simple cathartic sharing for the participants. The latest climate change talks that took place in Bonn, Germany in May and June of 2010 demonstrate the ongoing unwillingness to enact meaningful measures to confront climate change. These talks resulted in a text that was little more than the Copenhagen accord and ignored all measures proposed at Cochabamba. Frustrated with this outcome, Bolivia's representative to the UN, Pablo Solón pleaded with the president of the talks to listen to those who were actually facing the threats of climate change and create a framework that was more than just a "Copenhagen Plus" (Solón, 2010).

Conclusion

The Cochabamba Declaration has provided a foundation but mechanisms must be developed to carry its proposals forward. Unfortunately, the power to implement such mechanisms eludes the majority of the participants at the conference. Although the

Cochabamba Declaration recognizes the power imbalances with regards to climate change, it is hampered by that very same contradiction; those who are most affected are precisely those who have the least capacity to effect change. After years of being told to think globally and act locally, it is time the Global North took up its responsibility to act globally as well. It is time that citizens of the Global North demand the same seriousness of their governments as was seen among participants at Cochabamba. The Cochabamba Declaration, while still imperfect, is the result of a collaborative process and represents the most democratic document on climate change to date. The cry at Cochabamba was for a revolution in all spaces, places and on all fronts — it's time for citizens of the Global North to take part.

Figure 9: Participants celebrate at the closing of the conference in the stadium in Cochabamba (photo by Raul Burbano).

* * *

Sonja Killoran-McKibbin holds a master's degree in Planning and Political Economy from the Universidad Mayor de San Andrés in La Paz, Bolivia. She is currently a PhD student in the Faculty of Environmental Studies at York University in Toronto, Canada.

REFERENCES

CONAMAQ. "Declaración de la Mesa 18 : Derechos colectivos y derechos de la Madre Tierra," Tiquipaya.

Kranjc, Anita, 2006. "In Defence of the Environmental State: NGO Strategies to Resist the Commodification of Nature," pp. 187-202. In Gordon Laxer and Dennis Soron, editors, *Not for Sale: Decommodifying Public Life*. Peterborough: Broadview Press.

McAfee, Kathleen, 1999. "Selling nature to save it? Biodiversity and green developmentalism," *Environment and Planning D: Society and Space*, 17, 2, 133-154.

Ministerio de Planificación, 2006. *Plan Nacional de Desarrollo: 2006-2011*. La Paz.

Morales Ayma, Evo, 2010. "Call to the World Conference on Climate Change and the Rights of Mother Earth." http://cmpcc.org/2010/01/05/call/, accessed 25 May 2010.

People's Agreement, 2010. "People's Agreement on Climate Change and the Rights of the Mother Earth," April 22, Cochabamba, Bolivia. http://cmpcc.org/acuerdo-de-los-pueblos/, accessed 12 May 2010.

Prada, Raul, 2010. *Vivir Bien: Proyecto Civilizatorio Alternativo al Capitalismo*. La Paz: Ministerio de Planificacion.

Solón, Pablo, 2010. "Speech at the UN Climate Change Talks in Bonn." http://pwccc.wordpress.com/2010/06/11/bolivias-statement-on-new-texts-at-climate-negotiations, accessed 12 June 2010.

Stefanoni, Pablo, 2010. "¿Adónde nos lleva el Pachamamismo?" *Rebelión*. Page 7, 28 April. http://www.rebelion.org/noticia.php?id=104803, accessed 12 May 2010.

COP15 in an Uneven World
Contradiction and crisis at the United Nations Framework Convention on Climate Change

Introduction

When one thinks of the United Nations, images of democracy, compromise, and world leaders tackling difficult collective action problems come to mind. UN decisions may include the cooperative reduction of nuclear armaments, declarations of human rights, or peacekeeping operations. When it comes to climate change, the United Nations is also the highest decision-making body that nations turn to in order to come to a collective agreement on how to reduce global greenhouse gas (GHG) emissions. Climate change policy is negotiated at the global level mainly because the climate system's dynamics are globally integrated. The sources of GHG emissions can emerge anywhere on the globe, and may change climate conditions anywhere. Complicating matters, the sources of anthropogenic GHGs include regions, nations, localities, individuals, firms, and multiple activities. Due to these characteristics, climate change is considered a global collective action problem. In 1990, the UN General Assembly passed a resolution to formally launch negotiations towards an international climate change agreement and on May 9, 1992, the United Nation Framework Convention on Climate Change (UNFCCC) was adopted (IPIECA,

2008: 2). Currently, the Convention has been signed by 191 nations. The annual Conference of the Parties (COP) is the highest decision making authority of the UNFCCC. The COP is mandated to review the implementation of the Convention, to adopt decisions to further the Convention's rules, and to negotiate new commitments (IPIECA, 2008: 4). The Kyoto Protocol of 1997 was an outgrowth of this process and served to strengthen the Convention by setting binding targets on GHG emissions.

The fifteenth COP took place in Copenhagen from December 7-18, 2009, and was arguably one of the most important events in climate politics in the past decade. Governments, interest groups, and the media anticipated that COP15 would lead to a post-Kyoto international agreement. COP15 invited record public attention, non-governmental organization (NGO) participation, and political mobilization. Despite the high expectations of many observers, by all accounts, COP15 was a dismal failure. No binding accord was signed between nations, sending world leaders home with no new emission reduction targets. Civil society, which included thousands of invited NGO delegates, was literally locked out of the conference center in sub-zero conditions. Finally, Copenhagen itself was turned into a virtual police state where violence and police brutality were arbitrarily used against thousands of peaceful protestors. The questions that this chapter seeks to answer are: Why was COP15 a failure? And what are the consequences of COP15 for the future of international climate politics?

UNFCCC multilateralism

Commonly, the UNFCCC, COP, and the Kyoto Protocol are discussed as exemplars of international cooperation, conjuring up an image of multiple countries working in harmony towards resolving the single collective action problem of climate change. Central to the UNFCCC is the idea that cooperation among interested parties, including states, corporations, and civil society, can result in policies that resolve global warming while also maintaining economic prosperity. It is assumed that all parties share the common goal of atmospheric protection and that conventional science is the appropriate basis for environmental policy (Glover 2006: 6). These assumptions serve to situate climate change as a global environmental management problem.

The theoretical ideal and pragmatic means for international cooperation on climate change has been democratic pluralism. This involves individuals and groups competing, freely and openly, towards their own political ends through formal political processes (Dickerson & Flanagan, 1998). The annual COP serves as a space for nations to evaluate, negotiate, and improve their commitments within the Convention. The COP involves heads of state, national delegations, coordinated groups[1], and observer organizations. Any organization qualified in matters covered by the Convention and upon request, is admitted by the secretariat. The secretariat prides itself on the exceptional level of participation that observers enjoy. The basic assumption underlying the COP is that interest groups can lobby governments on an equal and level playing field to further their own political ends. The COP actively involves NGOs which typically attend sessions to observe and exchange views with other participants (IPIECA, 2008: 9). This involvement allows a wide range of groups to bring their experience, expertise, and perspectives into the climate negotiations. Scholars have argued that the influence of observers allows for innovative compromises that advance an international consensus. The array of observer NGOs typically attending the COP include business and industry groups, environmental groups, Indigenous groups, local governments, research groups, trade unions, women's groups, and youth groups. Agreements with UNFCCC are reached through negotiations and equal voting privileges between nations, while the powers of the UNFCCC over state sovereignty remain limited. In sum, the COP is mandated to operate according to the ideals of pluralism, liberal democracy, inclusion, and compromise.

Contradictions within the UNFCCC

Despite the UNFCCC's mandate for pluralist democratic engagement, the climate negotiation process has run into a number of contradictions throughout its short history. Understanding climate change politics in the international context requires an appreciation of the way in which political power is exercised by different groups in pursuit of their goals (Newell, 2000: 1). The claim that the UNFCCC represents a fair, equitable, and socially optimal approach ignores the uneven power relations between states or the uneven power between non-state actors. Political economy is valuable for analyzing the international interactions of states, corporations, and capital in shaping the international climate change negotiations.

The UNFCCC in an uneven world system

States do not enter the UNFCCC negotiations as equal players. The world economic system is characterized by asymmetrical power relations between nations, with the core industrialized countries holding a majority share of global wealth and contributing disproportionately to global GHG emissions.[2] The largest and most powerful states in the global economy are sustained by the use of cheap and readily available fossil fuel energy (Newell, 2000: 8). The particularities of the availability of fossil energy in different countries, helps explain differential bargaining positions and the dynamics of climate policy making (Newell & Paterson, 1998). Compounding this uneven pattern of consumption and wealth is the likelihood that developing nations are expected to experience the highest incidences of climate impacts and vulnerability. These asymmetrical relations have generated differential visions regarding the allocation of responsibility for climate change response. Systematic inequality has served to engender non-cooperation and distrust in international climate negotiations. Consequently, nations with the most power eschew responsibility for climate change while those with the greatest vulnerability to climate change have little bargaining power while carrying the heavy social costs (O'Hara, 2009: 230). Moreover, global resentment has been garnished as the South expects the North to reduce their consumption, while the North expects the South to make adjustments to reduce extreme events and GHG emissions (O'Hara, 2009: 230). Divisions also exist between developing countries, which further complicates the negotiations. For example, the demands of the small islands differ significantly from emerging economies. Distrust generated by inequality, power differentials, and divergent world views were major roadblocks to creating a post-Kyoto accord at COP15.

> **Systematic inequality has served to engender non-cooperation and distrust in international climate negotiations.**

Corporate power and the UNFCCC

Furthermore, not all NGOs share equal influence over the negotiation process. Typically, state positions are swayed by lobbies. One aspect of the COP that democratic pluralism ignores is the fact that corporations enjoy privileged access to and influence over state entities engaged in the negotiations (Newell & Paterson, 1998: 680). Since the inception of the UNFCCC, the fossil fuel lobby has been influential in climate negotiations. Why is it that the industry with vested interest in a weak climate agreement has historically played a privileged role in the climate negotiations? And what are the implications of this influence?

Fossil fuel lobbies have overtly acted to sway national, regional, and global responses to climate change in favor of their interests. The most famous example of corporate influence over the climate negotiations is the Global Climate Coalition, which describes itself as the 'leading business voice on climate change', and includes over 55 business associations and companies such as the American Petroleum Institute, DuPont, Ford, General Motors, Texaco, Chevron, and Shell (Newell & Paterson, 1998: 682-683). In a campaign from 1988-1999, the Coalition spent over U.S. $63 million on climate skepticism to change U.S. electoral opinion, 'anti-Kyoto Protocol' advertising, and on financial contributions to politicians opposed to a U.S. carbon tax (Glover, 2006). The UNFCCC process assumes that states and markets are separate, and that states make autonomous decisions. However, in capitalist societies the state plays a crucial function in maintaining the conditions of capital accumulation, which since the 20th century have relied on fossil fuel energy. The primary role of fossil fuels in both economic growth and global warming implies that industrialized economies are directly threatened by emission limits. The deep-seated contradiction between ecological limits and economic growth based on fossil fuels plagues the climate negotiations. Corporate lobbies limit the scope of state response to climate change thereby permitting continued consumption of fossil fuels in the industrialized core while also ensuring that international climate agreements open market opportunities. Lobbies representing fossil fuel companies have had their positions adopted by national governments and incorporated into the UNFCCC (Newell & Paterson, 1998: 682-683). At COP15, fossil fuel lobbies used their domestic influence to sway state negotiation positions. For example, the U.S. oil and gas industry spent $35 million, electric utilities spent $20 million, and the coal mining industry spent $3.4 million on political contributions to state representatives in order to protect energy industry interests at the COP15 negotiations (Open Secrets, 2010).

In addition, the UNFCCC structure includes a separate consultative mechanism for industry groups, giving the corporate lobby special access to the secretariat. At COP15, the UNFCCC held a Business Day, a consultative event, which featured over 40 speakers from the private sector and over 400 industry participants (Davenport, et al., 2009). Thousands of corporate lobbyists attended COP15. Among them were the World Business Council on Sustainable Development (WBCSD) (230 delegates), the International Emissions Trading Association (486 delegates), and the International Chamber of Commerce (ICC) (136 delegates) (UNFCCC, 2009). These groups have been criticized for their role in sabotaging the climate negotiations, while supporting false market solutions to climate change such as carbon capture and storage, and reduced emissions through decreased deforestation (REDD). The aim of the powerful corporate lobby at the COP15 was to influence the climate negotiation process towards protecting the interests of capital. The contradictory consequence of these on-going dynamics has been climate change solutions that favour the interests of capital under the guise of climate response.

Crisis at COP15

It is arguable that these contradictions reached a tipping point at COP15. Six months after, the executive secretary conceded that "Copenhagen was a pretty horrible conference" (van den Bosch, 2010). In fact, COP15 was a multifaceted failure in international politics and a moment of crisis for the UNFCCC. The failure of states to reach an accord, the NGO lockout, and the use of violence, all combined to create a crisis for the international climate process.

Non-binding Copenhagen Accord

COP15 failed to achieve a binding legal agreement for the post-Kyoto period. Although the Copenhagen Accord endorses the continuation of the Kyoto Protocol, it is not a legally binding accord, and it does not commit countries to new GHG reduction targets. The Accord offers to mobilize U.S. $100 billion to developing nations, but only in 2020 with no indication of how to do so. Moreover, the Accord was drafted by only five countries, the United States, China, India, South Africa, and Brazil, and emerged from an exclusive meeting outside of the COP. As such, the Accord was rejected by a number of states on the grounds of 'undemocratic' procedure. The status of the Accord and its legal implications remain unclear.

The NGO lockout

The official lockout of civil society from COP15 was a turning point for international climate politics. By December 14, 45,000 official delegates descended on COP15 to participate as observers. This historical turnout proved to be a serious challenge for the United Nations. Logistically, the conference site

Figure 1: Delegates wait for entry, December 14 (photo by Janina Schan).

could hold only 15,000 people, leaving 30,000 delegates out in the cold. Delegates were stranded, day after day, with no entry in sight (Figure 1).

Outraged with the shutout of their participation, NGO delegates protested. On December 16, locked-out NGOs joined a

Figure 2: Climate Justice Protest, December 16 (photo by author).

street march while NGOs inside COP15 protested (Figure 2). Eventually, a few hundred NGO delegates walked out of the conference to join their peers in the street. The delegates were met with police brutality and arbitrary arrest. Amid the chaos, the president of the UNFCCC resigned. The next morning the UNFCCC unilaterally decided to formally lock out all 15,000 NGO delegates from COP15. Thousands of invited participants were officially blocked from the multilateral climate process, marking the end of NGO participation within the UNFCCC.

The sudden absence of civil society groups was described as conspicuous and sobering (Hack, 2009). According to the UNFC-CC, the problems associated with including civil society were logistical, "stretching the organizational capability of the secretariat as never before." However, NGOs refused to accept this rhetoric and argued that the lockout was an effort to shut out voices that disagreed with the politics and weak proposals of the UNFC-CC. For NGOs whose UN status was deactivated, and whose members confronted police violence and arrests, COP15 was a moment of crisis, and decisive failure in UN multilateralism.

On December 17, NGO delegates held an emergency meeting at Klimaforum to discuss their future engagement with the UNFCCC. The sentiment was that the UNFCCC and national leaders did not want to include input from civil society and instead sided with transnational capital. Currently, it remains uncertain if the COP will ever include NGOs directly again. COP16 proposes to have two separate venues, one for national delegates and the other for NGOs. A public outcry against this new exclusionary structure has ensued.

Protests, violence, and containment of civil society

The most significant failure of COP15 was the use of violence to silence dissenting voices. The number of demonstrations in the streets outside of COP15 was unprecedented in UNFCCC history. Protesters included environmental activists, NGOs, and UN delegates. As the conference proceeded, it became clear that maintaining repressive control by police force was a priority for the UN and Denmark.

Prior to COP15, Denmark passed a law permitting the police to make pre-emptive arrests and detain anyone for up to 12 hours.[3] On December 9, police raided protestor accommodations, and detained 200 people. On December 12, over 100,000[4] climate justice protesters peacefully demonstrated, and police responded by pre-emptively detaining 968 protestors. By December 16 tensions between civil society and the UNFCCC escalated, and 35,000 protesters, including locked-out NGOs and delegates unpleased with the talks, marched to the gates of the conference

Figure 3: Police enforcements at COP15, December 16 (photo by author).

demanding to be heard. They were met by 9,000 police officers who arbitrarily used pepper spray, batons, and brutal force to contain them (Figure 3).

In light of these events, the executive director of Greenpeace UK wrote "The city of Copenhagen is a crime scene tonight … it is now evident that beating global warming will require a radically different model of politics than the one on display here in Copenhagen" (BBC, 2009). Prominent NGOs and activists argued that an alternative politics to the UNFCCC is necessary, and hope turned to the first World People's Summit on Climate Change.

Alternative visions

The first World People's Summit on Climate Change was an international gathering of 30,000 people that took place in Cochabamba from April 19-22, 2010 (*Green Left*, 2010). The conference was a response by civil society and several governments to the failure of COP15 and the UNFCCC process. The summit addressed the structural causes of climate change and the uneven pattern of development, production, consumption, and ecological degradation characteristic to global capitalism. The summit overtly critiqued the UNFCCC process and suggested alternatives such as a global referendum on climate proposals at the UNFCCC along with the establishment of an international climate justice tribunal to hold countries legally accountable to their Kyoto commitments. The conference produced a Universal Declaration of Rights of Mother Earth which was submitted to the UNFCCC. Finally, the conference established the People's Accord, which rejects the Copenhagen Accord, and places the onus for climate deadlock on corporations and governments in developed countries. The People's Accord proposes deep reforms to the UNFCCC, including state compliance and full consultation and participation for Indigenous peoples at international climate negotiations. Proponents of the Accord argue that it moves away from the UNFCCC towards a 'real' solution to climate change founded on resistance, a rejection of capitalism, and a revival of 'Indigenous' environmental values.

Despite the need for an 'alternative' to the UNFCCC, it is important to note that accepting the movement as 'liberatory'[5] may not revise environmental discourses in a direction that favors marginalized people. Calling to replace the capitalist system with socialist alternatives may not necessarily address emissions, climate change, or ensuing climate vulnerability. Moreover, it is important to note that the movement may also impose an ideology that falsely romanticizes the Indigenous relationships with nature. Most importantly, the movement may perpetuate a false vision of nature as existing apart from humanity and as such can result in the failure of 'alternative visions' to address modernity as the core cause of climate change and its politics.

Conclusion

In conclusion, the international process for negotiating climate change solutions has historically been less inclusionary, equitable, and multilateral than commonly assumed. The lens of political economy calls into question the myth of UNFCCC multilateralism by looking at the multifaceted failures of COP15. The UNFCCC is characterized by contradictions that reached a point of crisis at COP15. The failure of COP15 to reach a binding agreement, while locking out NGOs and using repressive force against civil society, marked a turning point in climate politics. In response, a counter approach, embodied in the People's Accord has emerged. However, the liberatory potential of this the people's climate movement for transforming human consciousness towards nature, and ultimately addressing the challenge of climate change, requires further inquiry.

* * *

Jacqueline Medalye is a PhD candidate in Political Science at York University. Her current research focuses on international climate change policy and its implications for development in the Global South.

ENDNOTES

[1] i.e., the G77, ASIS, and OPEC.

[2] The absolute emissions of India and China continue to rise however per capita emissions remain lower than the core industrialized countries.

[3] Police were allowed to detain anyone whom they suspect might break the law in the near future.

[4] Estimates vary from 25,000- 100,000. Even at low estimates, the scale was unprecedented.

[5] 'liberatory' is used here in Forsyth's sense.

REFERENCES

Bernstein, Steven, 2001. *The Compromise of Liberal Environmentalism*. West Sussex: Columbia University Press.

Green Left, 2010. "Bolivia's world climate summit a breakthrough," 1 May. http://www.greenleft.org.au/node/43891 accessed 1 May 2010.

BBC, 2009. "Copenhagen deal reaction in quotes," 19 December. http://news.bbc.co.uk/1/hi/sci/tech/8421910.stm, accessed 9 June 2010.

Davenport, Deborah, Morgera, Elisa, and Wagner, Lynn, 2009. "Summary of Copenhagen Business Day," *IISD Bulletin*, 160, 2. http://www.iisd.ca/climate/cop15/bd/html/ymbvol160num2e.html, accessed 24 May 2010.

Dickerson, Mark O., and Flanagan, Thomas, 1998. *An Introduction to Government and Politics. 5th Edition*. Scarborough: ITP Nelson.

Forsyth, Tim, 2004. "Industrial pollution and Social Movements in Thailand," pp. 422-438. In Peet, Richard and Watts, Michael, editors, 2004, *Liberation Ecologies: Environment, Development, and Social Movement.* 2nd Edition. London: Routledge.

Glover, Leigh, 2006. *Postmodern Climate Change*. London: Routledge.

Hack, Tobin, 2009. "Lockout: Civil Society Groups Denied Access to COP15," *Big Think*, December 19, http://bigthink.com/ideas/17943, accessed 8 June 2010.

Musumali, Abel, 2010, *Summary Report UNFCCC Bonn 1 Meeting*. http://www.ayicc.net/wp-content/uploads/2010/04/BONN-1-MEETING.pdf, accessed 7 June 2010.

Newell, Peter, 2000. *Climate for Change*. Cambridge: Cambridge University Press.

Newell, Peter, and Paterson, Matthew, 1998. "A Climate for Business: Global Warming, the State, and Capital," *Review of International Political Economy*, 5, 4 (December), 679-703.

O'Hara, Philip Anthony, 2009. "Political Economy of Climate Change, Ecological Destruction, and Uneven Development," *Ecological Economics*, 6, 2 (December), 223-234.

Open Secrets Database, 2010. http://www.opensecrets.org/, accessed on 13 June 2010.

The United Nations Framework Convention on Climate Change and its Kyoto Protocol: A Guide to Climate Change Negotiations, 2010.

http://ipieca.org/activities/climate_change/downloads/publications/unfccc_g uide.pdf, accessed 24 May 2010.

UNFCCC, 2010. *Communication Letter to Observers*, 23 February.

UNFCCC, 2009. *COP15 - FCCC/CP/2009/MISC.1*, 7 December.

van den Bosch, Servaas, 2010. "Climate Change: Restoring Trust after 'Horrible' Copenhagen Conference", *Inter Press Service*, 7 June. http://www.alertnet.org/thenews/newsdesk/ips/873faf5f335bb64e455c1321c 102fa9c.htm, accessed 7 June 2010.

World People's Conference on Climate Change and the Rights of Mother Earth. http://pwccc.wordpress.com/2010/04/24/peoples-agreement/, accessed 9 June 2010.

AARON SAAD

Climate Change, Compelled Migration, and Global Social Justice

In the lead-up negotiations to the fifteenth United Nations Conference on climate change in Copenhagen, a small advocacy group succeeded in having two short entries included in the draft many hoped would be finalized as a new, binding global framework for addressing climate change (DaSilva, 2009). The entries concerned those people, overwhelmingly in the Global South, who will be displaced or compelled to migrate in response to the effects of global warming — often called *climate refugees, environmental refugees*, or *environmental migrants*. Had the draft been adopted, it would have stood as a unique instance of recognition at the global level of their growing plight. However, those two entries were dropped when the entire text was replaced by the Copenhagen Accord, a non-binding agreement that included no timetable or method for reducing emissions and nothing on the matter of climate change and compelled migration.

The matter of climate change and migration resurfaced at the World People's Conference on Climate Change and the Rights of Mother Earth in Cochabamba, Bolivia mere months later — and the contrast was stark. Here, an entire working group was devoted to the issue, and its final conclusions contained several remarkable features, including the demand that capitalist countries accept responsibility for climate change-induced displacement by being the main contributors to a special global fund for climate migrants (Working Group 6, 2010).

At the time of writing, the 2010 Bonn Climate Talks had recently concluded, and another small entry on climate-related displacement and migration was included in a new negotiating text (Ad Hoc Working Group, 2010: 15). But like the two before, there is something missing: the states most responsible for causing climate change will have no binding responsibility towards people compelled to migrate in response. It will likely be some time before this changes. According to Dr Koko Warner, one member of the group negotiating for recognition of these migrants and displacees in the international system, identifying liable parties is a sure way to stall or even kill progress on the issue, and risks its ejection from negotiations (Warner, 2010).

This, then, is the chilly climate confronting advocates seeking just solutions on the issue of climate change-induced migration.

This, then, is the chilly climate confronting advocates seeking just solutions on the issue of climate change-induced migration. Before moving on, it is worth recalling, briefly, what some of the effects of climate change will be if serious action to reduce greenhouse gas emissions is not taken. Sea levels are projected to rise by 0.6m to 2m (range built from Pfeffer, Harper, & O'Neel, 2008; Vermeer & Rahmstorf, 2009; Jevrejeva, Moore, & Grinsted, 2010) by the end of this century, inundating low-lying coastal areas and small island nations. In the short term, rapid glacier melt will increase risks of flooding during wet months and, over the long term, decrease the availability of fresh water in glacier-fed basins during dry seasons as warming alters glacial volume, storage capacity, and replenishment rates (Stern, 2007: 76-79); water stresses will thereby increase sharply for the "more than one-sixth of the Earth's population relying on melt water from glaciers and seasonal snow packs for their water supply" (Rosenzweig et al., 2007: 187). In the subtropics, precipitation is projected to fall in intense events with high runoff interspersed with longer dry periods, causing both flooding and drought

(Meehl et al., 2007: 782). Dry areas are expected to become even drier; 1 to 4 billion people in Africa, the Middle East, Southern Europe, and Central and South America could experience water shortages (Stern, 2007: 74-76). By one estimate, "the proportion of the land surface in extreme drought, globally, is predicted to increase [...] from 1-3% for the present day to 30% by the 2090s" (Rosenzweig et al., 2007: 187). Extreme weather events such as tropical cyclones will increase in intensity and possibly frequency (Meehl et al., 2007: 788), causing massive destruction and casualties in vulnerable coastal regions. And these are just some of the projections. That all of this holds extremely dire implications for food production, water availability, individual and community livelihoods, health, habitation, and security should be clear — as should be the possibility that people will choose or be forced to migrate in response. Indeed, it is widely projected that the number of people displaced will surpass anything the world has yet seen.

This chapter sets out to introduce readers to the matter of climate change and induced migration, an emerging field of research and activism, and a crucial dimension of global warming that has received too little attention. It looks first into the problem of defining the environmentally displaced in the international system and how this is complicated by the complex intersection of the effects of global warming with existing socioeconomic factors. It then presents Bangladesh as a case study before taking up the issue of proposed solutions. The chapter concludes by looking at climate change and migration through the lens of global social justice.

Definitions and the multicausality of environmental migration

Though multiple terms have been proposed to refer to those who will choose, feel compelled, or be forced to leave their homes due to the effects of climate change, none have yet been broadly accepted. Of those suggested, *environmental refugee* and *climate refugee* have gained the most currency in the media and among the public, but the terms are extremely problematic for several reasons.

The first is that they tend to suggest that the displaced are victims of sudden deleterious environmental change who have no choice but to leave their homes. This oversimplifies a vast range of possibilities of how human migration and climate change intersect. Not all of the effects of climate change will manifest as sudden devastating events, like extreme cyclones that capture headlines and demand immediate response. Many, like desertification or erosion, will happen over a longer time-frame leaving affected populations more time to try to adapt — in fact migration is understood to be a strategy of adaptation. The term *refugee* is not the best fit for all cases and should per-haps be reserved for only a certain subset.

Furthermore, a crucial point that has emerged in the litera-ture (e.g., EACH-FOR, 2009: 71-72) is that migration decisions are multicausal, involving not only environmental issues, but also multiple socioeconomic ones. Migrants experience both "push" factors at place of origin (e.g., environmental degradation, conflict, lack of employment) and "pull" factors in destination areas (e.g., better economic opportunities, safer political atmos-phere) in their choice to move. So not only is the term *refugee* inaccurate, but because the environment is difficult to disaggre-gate from a complex, multicausal process, even use of the word *environmental* is open to criticism.

Another reason that the aforementioned terms are problemat-ic is that no matter how severe the environmental triggers, those displaced by the effects of global warming will not qualify as refugees under international law and are therefore ineligible for protection from existing agencies in the international system. This is because the word *refugee* has a specific legal meaning pro-vided by the United Nations Refugee Convention which identifies refugees as those people outside of their country of nationality whose displacement is due to well-founded fears of persecution for reasons of race, religion, or political or social membership.

Two issues arise. One is that the environment does not *perse-cute* people. The other is that — in contrast to popular notions of millions of environmentally displaced people clamouring for entry into the Global North — the vast majority of environmen-tally induced migration is projected to occur *within* the Global South, not outside of it, because of significant barriers to move-ment such as the high cost of migrating to distant destinations. The Office of the United Nations High Commissioner for Refugees has so far rejected proposals to change the current legal definition of refugee for fear that this may undermine existing refugee protections (Guterres, 2008: 7).

Bearing these and other criticisms in mind, some (e.g., Bates, 2002) have proposed looking at environmental migration as taking place along a continuum, and building from that a system of typologies that acknowledges that environmental signals can vary in strength and duration as well as intersect with other socioeconomic factors. However, nothing like this has achieved legal status.

It is widely understood that if allowed to progress along current trajectories, climate change will displace unprecedented numbers of people.

In sum, while the matter of terminology remains unsettled, the following is clear. Except in the most severe of rapid onset disasters, environmental change is but one factor intersecting with other socioeconomic factors that may or may not lead to migration. As climate change grows more acute, however, the environmental signal will grow stronger in an increasing number of people's ultimate decision to move. But just how many remains unknown. Though numbers are given frequently — the most common being 200 million by 2050 — no reliable ones currently exist. Nevertheless, it is widely understood that if allowed to progress along current trajectories, climate change will displace unprecedented numbers of people.

A glimpse of the future?

Before turning to the issue of solutions, it is important to look at a case showing how climate change and migration could play out. In 2005, the *Guardian* reported that the people of Papua New Guinea's Carteret Islands would need to be permanently relocated due to rising sea levels (Vidal, 2005). They are often called the world's "first climate refugees." While stories of people displaced by the effects of climate change have begun to appear more regularly ever since (e.g., Knight, 2009; Collectif Argos, 2010; rockhopper.tv, 2009), the attempt to find an uncontroversial case of climate change-related migration remains challenging. 2009 saw the publication of the important EACH-FOR study on environmental change and migration which produced one of the strongest cases yet for current migration compelled by climate change.

Bangladesh has been described as "the ground zero of global warming" (Atiq Rahman quoted in Dyer, 2008: 59). As a deltaic country, most of it stands less than 12 metres above sea level — and much of it actually below. Normally, during the monsoon season, one-third of the country becomes flooded — a phenomenon locals have designed ways of adapting to. However, with increased glacier melt, much more water now flows into Bangladesh's rivers, while sea-level rise slows its outward flow; in fact, in the delta, instead of freshwater going to sea, seawater now flows inwards, destroying crops. All of this causes the annual floods to increase gradually in volume and duration and, in 2007, they caused 3,363 casualties, affected 10 million people, and reduced agricultural yields by 13%. This increased flooding has also severely exacerbated the riverbank erosions that made 135,632 families homeless in the five years preceding the EACH-FOR study. Furthermore, the country is at high risk of cyclones (Poncelet, 2009: 6); more than five million Bangladeshis live in areas highly vulnerable to cyclones and storm surges (Warner, Ehrhart, de Sherbinin, Adamo, & Chai-Onn, 2009: 13). In November 2007, as the rehabilitation response to flooding in the previous July through August was still underway, Sidr, a 240km/h cyclone, struck, affecting 8.7 million people.

To those affected by these phenomena, migration has become one of the best strategies to adapt to and survive these impacts. But it appears to be the best only among very limited options — and in some cases is not an option at all. First, many lack the financial means to migrate, despite wanting to. Others who have made the attempt were robbed or sent by traffickers to Indian sweatshops, where they worked as slaves. Women — especially the unmarried and widowed (many due to their husbands' work in deadly flood or storm zones) — and children experience special vulnerability to traffickers, some ending up as prostitutes or forced labourers. For people who do have the means to migrate, finding housing and employment is a serious issue. Many of those arriving in the slums of Dhaka are forced to compete with one another for physically demanding, poorly paid single-day contract jobs, and few ever find employment in the more steady garment or construction industries (Poncelet, 2009: 6-20). In sum, migration is not an option for all, and for those for whom it is, it can increase vulnerability. Migrants become targets for traffickers, face problems finding housing and employment, and add to an easily exploitable pool of desperate labour swelling already overpopulated city slums.

What Bangladesh provides is one possible glimpse into the future. If similar conditions are allowed to intensify and become more widely distributed throughout the Global South within a few decades, will the solutions that are currently being proposed be enough?

Solutions?

The reality is that there exist today no effective response mechanisms for people displaced for reasons of environmental change. There are, however, some recent notable trends. One has been the attempt to identify norms, practices, protocols, and so forth that already exist in the international system and that could be bulked up in existing humanitarian response mechanisms to fill protection gaps (e.g., Zetter, 2009). Related to this are proposals for new protocols on environmentally compelled migration (Biermann & Boas, 2010). A second development are proposals that developing countries incorporate plans for addressing environmentally compelled migration into their National Adaptation Programmes of Action, which are frameworks intended to identify priority areas for climate change adaptation (Martin, 2009).

A third recent trend has been to recognize that migration compelled by climate change has a positive side to it (e.g., Barnett & Webber, 2009). As more people migrate, one line of this argument goes, remittance networks will expand, and this will direct financial assistance towards underdeveloped regions. Moreover, as people emigrate, they relieve stress on environments suffering degradation from both climate change and human overuse. The solution that follows is to facilitate the movement and resettlement of people who will be displaced and their ability to send remittances home.

But there is the potential to overlook something important here. Once the barriers to migration are reduced or removed, dealing with the effects of the climatic changes their communities and nations had little role in causing becomes largely — perhaps entirely — the responsibility of the migrants. What should be a matter of global concern, where those most responsible and most able to provide funding and assistance play a lead role in ensuring a just response, can easily become a private matter for displaced individuals of poor nations to deal with.

Currently, the discourse on solutions is largely dominated by humanitarian, security (Dyer, 2008), and adaptation frameworks. Much like market approaches proposed for addressing climate change, what makes these so acceptable is that they identify a problem and suggest ways of responding to it that do not attempt to hold any parties responsible for causing that problem or bind them to a solution that truly redresses it. All of this should be seen in light of the fact that the people most vulnerable to and already affected by climate change — and therefore most likely to be pressured to migrate in response — have rarely been included in discussions on this matter; indeed, one of the most crucial knowledge gaps in this entire field is what these people believe just and effective solutions are. In other words, the dominant frameworks have not been chosen by the people most affected by this issue, and happen to be the ones that absolve the countries that have caused it from having any binding role in resolving it. Virtually absent thus far from the acceptable discourse has been one framework that does demand that parties historically responsible for climate change redress its effects and that opens up space for vulnerable and affected communities to see realized the solutions they feel are most just and effective: global social justice.

The historical context of climate change-induced displacement

The first point to bear in mind when discussing climate change and displacement within a framework of global social justice is that the people who are most vulnerable to the effects of global warming — and therefore most likely to migrate or be displaced as a result — are those least responsible for it. Developing and least-developed countries, making up 80% of the world's population, are responsible today for 41% of global emissions and for only 23% of cumulative emissions since 1750 (Raupach et al., 2007: 10292). There has been some recognition of this in the literature, though it remains anathema in the realm of international political affairs.

Not only are these countries least responsible for climate change, but they are also least able to adapt to it due to poverty, debt, political instability, weak governance structures, and so on. Much of the reason for this lies in the fact that — and this is a point not acknowledged in the literature — the political, social, and economic realities of communities throughout the Global South have been shaped by a long history of injustices committed against them.

For some communities, it is a global history beginning with the twin demographic tragedies sparked by the European conquest of the Americas: the Atlantic slave trade, which stole millions of African lives and destabilized nations through to the eve of colonialism, and the displacement and virtual eradication of the Indigenous populations of the Americas who have been kept marginalized since, often violently. For others, this history begins with European colonialism, a period during which communities the world over were brought forcibly into the world economy to occupy some of its lowest tiers. Local industries were devastated and traditional livelihoods radically changed in order for the colonized to supply cheap labour to export foodstuff and raw materials to the industrializing colonial powers and their offshoots — a pattern that continues to this day in much of the world, with transnational corporations replacing colonial enterprises.

The post-World War II period of decolonization offered few opportunities to break from this. Pursuing their Cold War aims, the United States and Soviet Union along with their allies toppled democratic governments throughout the Third World that were seeking independence through nationalist development programs and replaced them with allied dictatorships willing to keep their populations bound to the old colonial patterns. As neoliberal globalization took hold from the late 1970s onward, the Western economic powers used debt crises to impose structural adjustment programs on Third World economies through international financial institutions. These drastically reduced public programs in health, education, and food and opened these sectors to foreign private control — all while further deepening their international debt. Barriers to foreign goods and subsidized food were also eliminated, wiping out infant industries, sending millions of workers into the precarious informal sector, and driving subsistence farmers into poverty and hunger. The development paths taken by virtually all rich countries today were thereby effectively sealed off. It is because of all this that large swaths of the Global South will be unable to withstand the newest chapter of this history in which the world confronts climate change resulting primarily from the emissions of the rich, advanced industrial countries.

But recognizing these two matters of justice — the differentiated responsibility for climate change and the history that has made communities more vulnerable to its effects — means passing through a chilly climate. As discussed, attempts to make those countries most responsible for climate change legally part of a solution for the resulting migrations and displacement are strongly rejected by them, and can even endanger the inclusion of the matter in international negotiations.

Does seeking social justice, then, impede pragmatic solutions? It can, at least as long as rich and powerful states continue to negotiate from a position in which they see no reason to recognize their historical responsibility for this issue, or any reason to seriously address this matter at all. Developing countries are rarely in a position to change this. That role rests, then, with concerned populations throughout the Global North, who will need to compel their governments to accept responsibility for their historical role in both climate change and the underdevelopment of countries throughout the Global South — and to then redress this. Indeed, seeking social justice in this matter might even enhance solutions. It could very well mean the difference between the environmentally displaced receiving the bare minimum of what they need — if even that — under the currently acceptable non-binding frameworks, and receiving what they are owed.

* * *

Aaron Saad *earned his BA in Development Studies from the University of Calgary in 2004 and his MA in Globalization Studies at McMaster University in 2009. He is currently a PhD candidate in the Environmental Studies program at York University researching climate change and compelled human migration.*

REFERENCES

Ad hoc working group on long-term cooperative action under the convention, 2010. Advance draft of a revised text to facilitate negotiations among Parties, to be issued as an official document (FCCC/AWGLCA/2010/8) for consideration at the eleventh session of the AWG-LCA, 10 June. http://unfccc.int/files/meetings/ad_hoc_working_groups/lca/application/pdf/awg-lca_advance_draft_of_a_revised_text.pdf, accessed 14 June 2010.

Barnett, J. & Webber, M., 2009. *Accommodating migration to promote adaptation to climate change*. Commission on Climate Change and Development.

Bates, D. C., 2002. "Environmental refugees? Classifying human migrations caused by environmental change," *Population and Environment*, 23, 5, 465-477.

Biermann, F. & Boas, B., 2010. "Preparing for a warmer world: Towards a global governance system to protect climate refugees," *Global Environmental Politics*, 10, 1, 60-88.

Collectif Argos, 2010. *Climate Refugees*. Cambridge: MIT Press.

DaSilva, D., 2009. *Entries Added to Final Draft Agreement Before Copenhagen*. 27 November. http://www.towardsrecognition.org/2009/11/entries-added-to-final-draft-agreement-before-copenhagen/, accessed 14 June 2010.

Dyer, G., 2008. *Climate Wars*. Random House of Canada.

EACH-FOR, 2009. Synthesis Report (D.3.4.): EACH-FOR. http://www.each-for.eu/documents/EACH-FOR_Synthesis_Report_090515.pdf, accessed 14 June 14 2010.

Guterres, A., 2008. *Climate change, natural disasters and human displacement: a UNHCR perspective*. United Nations High Commissioner for Refugees.

Jevrejeva, S., Moore, J.C., & Grinsted, A., 2010. "How will sea level respond to changes in natural and anthropogenic forcings by 2100?" *Geophysical Research Letters*, 37, L07703, doi:10.1029/2010GL042947.

Knight. S., 2009. "The human tsunami," *Financial Times*, 19 June. www.ft.com/cms/s/2/bb6b0efc-5ad9-11de-8c14-00144feabdc0.html, accessed 14 June 2010.

Martin, S., 2009. "Managing environmentally induced migration," pp. 319-351. In F. Laczko & C. Aghazarm, editors, *Migration, Environment and Climate Change: Assessing the Evidence*. Geneva: International Organization for Migration.

Meehl, G.A., Stocker, T.F., Collins, W.D., Friedlingstein, P., Gaye, A.T., Gregory, J.M., Zhao, Z.-C., 2007. "Global Climate Projections," pp. 747-846. In Solomon, S., Qin, D., Manning, M., Chen, Z., Marquis, M., Averyt, K.B.,

Tignor, M., & Miller, H.L., editors, *Climate Change 2007: The Physical Science Basis. Contribution of Working Group I to the Fourth Assessment Report of the Intergovernmental Panel on Climate Change*. Cambridge: Cambridge University Press.

Poncelet, A., 2009. Bangladesh Case Study Report: The Land of Mad Rivers. EACH-FOR. http://www.each-for.eu/documents/CSR_Bangladesh_090126.pdf, accessed 14 June 2010.

Pfeffer, W. T., Harper, J.T., & O'Neel, S., 2008. "Kinematic constraints on glacier contributions to 21st-century sea-level rise," *Science*, 321, 1340-1343.

Raupach, M. R., Marland, G., Philippe, C., Le Quéré, C., Canadell, J. G., Klepper, G., & Field, C.B., 2007. "Global and regional drivers of accelerating CO2 emissions. *Proceedings of the National Academy of Sciences of the United States of America*," 104, 24, 10288-10293.

rockhopper.tv (Producer), 2009. *Hot Cities*. [Television documentary series]. http://www.rockhopper.tv/hotcities/index.html, accessed on 15 June 2010.

Rosenzweig C., Casassa, G., Karoly, D.J., Imeson, A., Liu, C., Menzel, A., Tryjanowski, P., 2007. "Assessment of observed changes and responses in natural and managed systems," pp. 79-131. In Parry, M.L., Canziani, F., Palutikof, J.P., van der Linden, P.J., & Hanson, C.E., editors, *Climate Change 2007: Impacts, Adaptation and Vulnerability. Contribution of Working Group II to the Fourth Assessment Report of the Intergovernmental Panel on Climate Change*. Cambridge: Cambridge University Press.

Stern, N., 2007. *The Economics of Climate Change: The Stern Review*. New York: Cambridge University Press.

Vermeer, M. & Rahmstorf, S., 2009. "Global sea level linked to global temperature," *Proceedings of the National Academy of Sciences of the United States of America*, 106, 51, 21527–21532.

Vidal, J., 2005. "Pacific Atlantis: first climate change refugees," *Guardian*, 5 November. http://www.guardian.co.uk/environment/2005/nov/25/ science.climatechange, accessed 14 June 2010.

Warner, K. 2010. Personal communication, 12 June.

Warner, K., Ehrhart, C., de Sherbinin, A., Adamo, S., & Chai-Onn, T., 2009. *In Search of Shelter*. CARE International.

Working Group 6 of the People's Conference on Climate Change and the Rights of Mother Earth, 2010. Final conclusions working group 6: Climate change and migrations. http://pwccc.wordpress.com/2010/04/30/final-conclusiones-working-group-6-climate-change-and-migrations/, accessed 14 June 2010.

Zetter, R., 2009. "The role of legal and normative frameworks for the protection of environmentally displaced people," pp. 385-441. In F. Laczko and C. Aghazarm, editors, *Migration, Environment and Climate Change: Assessing the Evidence*. Geneva: International Organization for Migration.

Part II

Chilly Climates

**STEPHANIE RUTHERFORD and
JOCELYN THORPE**

Framing Problems, Finding Solutions

How we understand a problem shapes the kind of solutions we can imagine and therefore work toward. Environmental problems are no exception. In this chapter, we describe how the idea of wilderness has resulted in environmentalist efforts to save wilderness areas from destruction. Yet attempts to save wilderness sometimes reinforce historical injustices — for example, the dispossession of Aboriginal peoples from their homelands — through which North American wilderness came into existence — and therefore do not provide a straightforward solution to the problem of resource depletion. Similarly, stories about climate change help us to understand the phenomenon and to consider how we should respond. But some narratives about climate change create problems of their own. We hope that by examining different ways of conceptualizing environmental concerns, we will open ourselves up to creative approaches to pressing issues such as climate change.

Temagami, Ontario, appears from the outside to be the epitome of Canadian wilderness. Indeed, hundreds of tourists descend upon Temagami each summer — and have done for over a century — to take advantage of the region's wilderness features: forested landscapes, rocky shorelines, lakes and rivers ideal for canoe travel, and fish and game aplenty (on the history of the Temagami region, see Hodgins & Benidickson, 1989; Thorpe, 2008a & 2008b). Robert Bateman has painted the Temagami pine trees. Margaret Atwood canoe-tripped through Temagami in order to protest the logging of the region. Bob Rae was arrested along with environmentalists when he helped blockade a Temagami logging road. These prominent Canadians aided in bringing national and international attention to the region in the late 1980s, when logging threatened to destroy what environmentalists called the last great pine wilderness (Bray & Thomson, 1990). Today, with some of the region classified as provincial parkland, Temagami's old-growth pine trees are one feature that attract tourists to the region.

As early as the late nineteenth century, visitors began to travel to Temagami in order to experience the Canadian wilderness, including its old trees. But in the past, even more so than in the present, relatively few people could afford a Temagami wilderness vacation, and it was largely middle- and upper-class white women and men who made their way from urban centres in Canada and the United States to experience what one visitor called "Nature's playground" (Another Wet Bob, 1899: 53). As with other places in Ontario, Temagami provided tourists with the opportunity to experience wilderness and to reflect upon the relative merits of "civilized" and "uncivilized" existence (Jasen, 1995). While the people who travel to Temagami today tend to avoid other people on their excursions, around the turn of the twentieth century tourists often considered interaction with local Aboriginal people to be a fundamental part of their wilderness vacations. Aboriginal people, like the forested landscape, appeared "uncivilized" to tourists, closer to wild nature than to Euro-Canadian culture (ibid.). Aboriginal people thus featured as part of the Temagami wilderness, not as people who owned or had any claim to the land that tourists visited.

The Teme-Augama Anishnabai, the Aboriginal people whom tourists encountered on their Temagami wilderness vacations, however, had a very different understanding of themselves and the region than did tourists. They knew themselves as the Teme-Augama Anishnabai, "the people of the deep water," and they conceptualized the land not as a wilderness, but as a homeland, "n'Daki Menan," or "our land." For the Teme-Augama Anishnabai, tourists were not visiting the Canadian wilderness, but rather were (sometimes welcome and other times unwelcome) guests to n'Daki Menan. Yet provincial and federal governments, like tourists, considered Temagami to be part of Ontario, Canada, and did not recognize the Teme-Augama Anishnabai's relationships with the region. Indeed, after almost one hundred years of struggling with provincial and federal government officials to have n'Daki Menan recognized as Teme-Augama Anishnabai territory, the First Nation took legal action in the 1970s to have n'Daki Menan recognized in Canadian law (McNab, 2009; Thorpe, 2008b). The First Nation was ultimately unsuccessful in its endeavour (although negotiations with the provincial and federal governments continue to the present), but the action itself reveals the Temagami wilderness to be a social and historical construction — and a colonial imposition — rather than a simple statement of fact.[1]

The idea that Aboriginal peoples, and indeed Indigenous peoples around the world, are "uncivilized" has lost at least some of its cultural salience since the dismantling of European empires and the rise of anti-colonial thinking.[2] In the Canadian context, for example, Prime Minister Stephen Harper has apologized for the policy of assimilation embedded in Indian residential schools. The assimilation of Aboriginal peoples into Euro-Canadian culture was long considered by the Canadian government to be foundational to the "civilization" of Aboriginal peoples. Yet now the government officially recognizes that this policy of assimilation not only caused great harm to Aboriginal individuals and cultures, but was also fundamentally wrong. The idea that Aboriginal peoples were uncivilized facilitated not only assimilationist policies like residential schools, but also the European takeover of Aboriginal lands, since, from a European perspective, "uncivilized" peoples did not own land, and so the lands inhabited by Aboriginal peoples could be seen as uninhabited, or wilderness, and open for European exploitation (see Harris, 2002). It is interesting to note that while the idea that there exists a hierarchy of civilizations in which Europeans are on the top and Indigenous peoples are on the bottom has been thoroughly debunked, the idea of wilderness has not been taken apart to the same extent (for notable exceptions, see Anderson, 2007; Braun, 2002), in spite of the fact that wilderness is a product of the same mindset that considered Aboriginal peoples to be uncivilized and therefore saw Aboriginal lands as unoccupied wilderness. But since the idea of Aboriginal savagery is recognized as a colonial fallacy, then wilderness too must be understood as a powerful colonial force, and as a myth.

Instead, wilderness has been held up by environmentalists as the ideal state of nature. In the late 1980s and early 1990s, environmental concern in Canada and the U.S. often centred on wilderness regions, for example Clayoquot Sound, British Columbia (Braun, 2002), Temagami, Ontario (Thorpe, 2008b), and areas around Portland, Oregon (Prudham, 2005). Controversies about whether these areas should be logged or left as wildlife habitats and tourist attractions received national and international media attention, and environmentalists garnered a great deal of public support for their attempts to save pristine nature. Yet environmentalist efforts, in reinforcing the idea of an empty wilderness — this time one that needed rescuing rather than visiting — often resulted in the repetition of historical erasures of Aboriginal peoples from the landscape (Braun, 2002; Thorpe, 2008b). Most recently, however, environmental concern seems to have shifted away from an emphasis on specific wilderness regions to focus on the issue of global environmental problems, with climate change being the most pressing. Climate change, like the industrial transformation of Native lands, is certainly an important environmental and social issue to which we must respond. But it also remains imperative to examine how the story of climate change is told, and therefore how solutions are envisioned.

In *Planet Dialectics: Explorations in Environment and Development* (1999), Wolfgang Sachs observes that the idea of global environmental problems, such as those put forward in climate change discourse, can work as a tool of exclusion. He argues that the images of earth from space generated by the 1968 Apollo 8 mission made possible the shift in Western thinking about environmental problems as local, regional or national in scope to global in character. The earth appeared in these photographs as small and vulnerable, a bounded sphere in need of protection. This view from above allowed for the construction of the earth as something that required management, an entity that only trained experts from the First World could understand

and control. Local knowledge about environmental problems, and the primarily Third-World local people who possessed such knowledge, became devalued within this emerging idea of global nature. This story, in light of the previous discussion of the erasure of Aboriginal claims to the land under colonialism, might seem familiar.

Sachs and others also express concern that the new understanding of environmental problems as global has created a "one-world discourse" in which all the earth's humans are connected through our intertwined ecological fate (Jasanoff, 2004; Sachs, 1999). In this view, all of us are equally endangered by environmental threats and all of us equally responsible for healing a planet in peril. But this way of understanding environmental problems once again erases the specificity of different peoples' relationships with the land, leaving no room to discuss issues such as Aboriginal land claims. These are the issues that also disappear when landscapes are considered to be part of the national wilderness rather than the territories of specific First Nations. In its assumption that we are all in the same boat, the one-world discourse also neglects that fact that we are neither all equally responsible for global environmental problems nor all likely to suffer the same consequences. For example, Canadian carbon emissions have increased by 26% rather than being reduced by 6%, as our commitment to the Kyoto Protocol promised, a commitment that Prime Minister Harper has now said the nation will abandon (Monbiot, 2009). Americans, who make up only 4% of the world's population, consume 25% of its resources. This picture is complicated, however, by statistics that show that African-Americans emit 20% fewer greenhouse gases per year than do Euro-Americans (Hoerner & Robinson, 2008). The one-world discourse glosses over the fact that we do not all inhabit one world of equal access to resources or equal threat of environmental risks. Indeed, if we adhere to the environmentalist call that "everyone is downstream," climate change makes it clear that, in the words of Jim Tarter, some of us "live more downstream than others" (2002: 213). The Indigenous peoples of the world, the citizens of island nations, and poor people — the most economically, politically, and environmentally vulnerable among us — feel the effects of climate change first and most strongly as problems like sea level rise, melting permafrost, poor air quality, and extreme weather events affect those with the least capacity to mitigate their effects.

In December 2009, we witnessed another round of negotiations among countries on targets for global emissions reductions. While several thousand delegates met in Copenhagen, Shishmaref, Alaska, home to over 600 Inupiaq individuals, threatened to fall into the sea because of permafrost melt as a result of rising temperatures (Carver, 2008). Indeed, the official meetings in Copenhagen and the events in Shishmaref represent two poles, if you will, in the debate on climate change: one a global, technocratic and managerial view, and the other representing the lived realities of our changing climate. As the global debate on how to manage climate change rages on, its impacts are felt on a very local level. There seems, in some sense, to be an inverse principle at work: those least responsible for the crisis are the most likely to bear its brunt. When we imagine climate change as only a global issue for which we all share equal responsibility, questions of justice are hidden.

This kind of understanding, however, is not the only option available when thinking about climate change. Indeed, scholar and environmental justice activist Giovanna Di Chiro (2003) offers us a way of thinking about all humans' shared connection to the earth that insists we pay attention to local people, places and problems. She discusses environmental justice groups that organize "toxic tours" in the U.S., in which visitors, guided by local residents, tour through poor and racialized communities and experience the environmental hazards faced by these communities. The tours stop at places such as playgrounds and housing developments, and "tourists" hear the stories of people who have developed cancer, respiratory diseases and immune deficiencies as a result of contamination and pollution in these neighbourhoods. Tourists viscerally experience the smells and sounds of communities that are host to waste incinerators, mountaintop removal coal operations, and oil refineries. Toxic tours generate awareness, exposing the largely white, middle-class tourists to the fact that poor and racialized communities live within the production and dumping grounds of industrial society. Tourists therefore see the local effects of global industrial processes. Simultaneously, toxic tours function to make connections between diverse communities: visitors see how their lifestyles and consumption patterns impact the local communities that bear a disproportionate environmental cost of these patterns. In this way, Di Chiro argues, political strategies like toxic tourism create a "globalized sense of place," which allows visitors and residents alike to see how their communities are tied together, and makes it possible for them to forge connections across difference to advocate for change (ibid.: 228-229).

This alternative vision of a common future is also present in the way some environmental justice groups have approached the discourse of climate change. The California Environmental Justice Movement, for example, shocked the broader environmental movement by releasing a declaration in February 2008 in opposition to cap-and-trade proposals to regulate carbon emissions. In their declaration, the group argued that these solutions to climate change, like the Kyoto Protocol, erase differences in responsibility for and impacts of pollution. These solutions entrench a business-as-usual approach to reducing emissions by maintaining reliance on fossil fuels while allowing corporations to pay to pollute. As the California Environmental Justice Movement (2009) points out, racialized people bear the burden of this particular brand of energy production, from extraction to waste disposal. They propose that to address the issue of climate change adequately, we must restructure the economy so that environmental and social justice are linked. They assert that "capturing energy from the wind, sun, ocean, and heat stored within the Earth's crust builds the health and self-reliance of people and our communities, protects the planet, creates jobs, and expands the global economy" (ibid.). In this way, the California Environmental Justice Movement calls for a more radical and sustained engagement with the problems associated with fossil-fuel dependence and climate change, one that pays attention to the differential impacts of climate change on people and places.

Similarly, Indigenous groups in both the U.S. and Canada have highlighted what they call the CO_2lonialism inherent in mainstream global solutions to climate change. *The Indigenous People's Guide: False Solutions to Climate Change* (2009) shows how current approaches to mitigating climate change serve to support the dispossession of Indigenous peoples in a "new land grab." They cite, for instance, the example of the Clean Development Mechanism of the Kyoto Protocol. This is a program through which countries in the Global North can invest in offsetting their emissions by supporting projects in the Global South like tree farms, which act as carbon sinks. Drawing on cases in Panama and Colombia, among others, *The Indigenous People's Guide* shows that Native peoples have been further dispossessed by this mechanism, as the people who live in these carbon sinks are relocated (ibid.). Indigenous groups also stand in opposition to carbon trading, "clean coal," and geo-engineering, which they, like the California Environmental Justice Movement, suggest serve only to reinforce ways of thinking that degrade the environment, Indigenous peoples and racialized people, thus reproducing the colonial thinking that dispossessed Aboriginal peoples during European colonialism. Indigenous groups propose instead the phasing out of fossil fuel-based energy, the promotion of Indigenous peoples' sovereignty, and the transition to "to sustainable models of production, consumption and development" (ibid.). The resistance of these groups reveals how we need to "look both ways" when we tackle global environmental problems like climate change, always considering how global issues are also local issues, and how environmental issues are also justice issues.

Climate change is a pressing environmental concern, one with serious impacts for all the earth's inhabitants. But accepting it simply as a global problem limits the solutions we can imagine, and risks reproducing inequities rooted in colonialism. By reframing the problem, it becomes possible to envision different responses, responses which might lead to more promising relationships among differently located humans and the changing climate we share, albeit unequally. Perhaps it is time to take the cue from environmental justice activists and Indigenous peoples who understand the land as part of the local and the global at the same time, and who call for us to remember our history as we approach the future.

* * *

Stephanie Rutherford *holds a Ph.D. in Environmental Studies from York University. She has taught at Macalester College where she received the Educator of the Year Award. She is currently teaching at Trent University in Peterborough.*

Jocelyn Thorpe *holds a Ph.D. in Environmental Studies from York University. She has held a post-doctoral fellowship in History at the University of British Columbia and is currently an Assistant Professor in Women's Studies at Memorial University.*

A slightly different version of this article originally appeared as "National Natures in a Globalized World: Climate Change, Power and the Erasure of the Local" volume 90.1 of the Dalhousie Review (Spring 2010): 127–138.

ENDNOTES

[1] On the historical creation of wilderness in the United States, see Cronon, 1996.

[2] We do not mean to suggest that this kind of racist thinking has disappeared entirely. Sadly, it has not.

REFERENCES

Anderson, Kay, 2007. *Race and the Crisis of Humanism*. London and New York: Routledge.

Another Wet Bob, 1899. "Temagaming," *Rod and Gun*,1, 3, 53.

Braun, Bruce, 2002. *The Intemperate Rainforest: Nature, Culture, and Power on Canada's West Coast*. Minneapolis: University of Minnesota Press.

Bray, Matt and Ashley Thomson, editors, 1990. T*emagami: A Debate on Wilderness*. Toronto and Oxford: Dundurn Press.

California Environmental Justice Movement, 2009. "The California Environmental Justice Movement's Declaration against the Use of Carbon Trading Schemes to Address Climate Change," http://www.ejmatters.org/declaration.html.

Craver, Amy, 2001. "Alaska Subsistence Lifestyles Face Challenging Climate," *Northwest Public Health,* Fall/Winter, 8–9.

Cronon, William, 1996. "The Trouble with Wilderness: Or, Getting Back to the Wrong Nature," pp. 69-90. In William Cronon, editor, *Uncommon Ground: Rethinking the Human Place in Nature*. New York: W.W. Norton & Co.

de Jong, Antoinette, 2008. "All at Sea," *Sun Herald*, 20 April, 42.

Di Chiro, Giovanna. 2003. "Beyond Ecoliberal 'Common Futures': Environmental Justice, Toxic Touring, and a Transcommunal Politics of Place," pp. 204-232. In Donald S. Moore, Jake Kosek and Anand Pandian, editors, *Race, Nature, and the Politics of Difference*. Durham and London: Duke University Press.

Harris, Cole, 2002. *Making Native Space: Colonialism, Resistance, and Reserves in British Columbia* Vancouver: University of British Columbia Press.

Hodgins Bruce W. and Jamie Benidickson, 1989. *The Temagami Experience: Recreation, Resources, and Aboriginal Rights in the Northern Ontario Wilderness*. Toronto: University of Toronto Press.

Hoerner, J. Andrew and Nia Robinson, 2008. *A Climate of Change: African Americans, Global Warming, and a Just Climate Policy in the U.S.* Oakland, CA: Environmental Justice and Climate Change Initiative and Redefining Progress.

Indigenous Environmental Network, 2009. *The Indigenous Peoples' Guide: False Solutions to Climate Change*. http://www.earthpeoples.org/CLIMATE_CHANGE/Indigenous_Peoples_Guide-E.pdf.

Jasanoff, Sheila, 2004. "Heaven and Earth: The Politics of Environmental Images," pp. 31-52. In Sheila Jasanoff and Marybeth Long Martello, editors, *Earthly Politics: Local and Global in Environmental Governance*. Cambridge, MA: MIT Press.

Jasen, Patricia, 1995. *Wild Things: Nature, Culture, and Tourism in Ontario*, 1790 –1914. Toronto: University of Toronto Press.

McNab, David T., 2009. *No Place for Fairness: Indigenous Land Rights and Policy in the Bear Island Case and Beyond*. Montreal and Kingston: McGill-Queen's University Press.

Monbiot, George, 2009. "Canada's Image Lies in Tatters: It is Now to Climate what Japan is to Whaling," *The Guardian* (November). http://www.guardian.co.uk/commentisfree/cif-green/2009/nov/30/canada-tar-sands-copenhagen-climate-deal.

Prudham, W. Scott, 2005. *Knock On Wood: Nature as Commodity in Douglas-Fir Country*. New York: Routledge.

Sachs, Wolfgang, 1999. *Planet Dialectics: Explorations in Environment and Development*. London: Zed Books.

Tarter, Jim, 2002. "Some Live More Downstream than Others: Cancer, Gender, and Environmental Justice," pp. 213-228. In Joni Adamson, Mei

Mei Evans, and Rachel Stein, editors, T*he Environmental Justice Reader: Politics, Poetics, and Pedagogy*. Tuscon: University of Arizona Press.

Thorpe, Jocelyn, 2008a. "To Visit and to Cut Down: Tourism, Forestry, and the Social Construction of Nature in Twentieth-Century Northeastern Ontario," *Journal of the Canadian Historical Association / Revue de la Société historique du Canada*, 19, 1, 331–357.

Thorpe, Jocelyn, 2008b. "Temagami's Tangled Wild: Race, Gender and the Making of Canadian Nature," PhD dissertation, York University.

NOËL STURGEON

Penguin Family Values
The nature of planetary environmental reproductive justice*

In 2005, a nature documentary entitled *The March of the Penguins* was a surprise hit, winning an Academy Award in 2006 for best documentary. The beautifully filmed story of the improbable but gorgeous Antarctic Emperor penguins and their incredible effort to produce and nurture their babies was a tale of terrific difficulties overcome through amazing persistence. In an interesting twist, and to the astonishment of the director, Luc Jacquet, right-wing fundamentalist Christians in the U.S. adopted the film as an inspiring example of monogamy, traditional Christian family values, and intelligent design. At around the same time, apparently unbeknownst to right-wing fundamentalist Christians, penguins had become a symbol of the naturalness of gay marriage.

Meanwhile, in other political and cultural discourses, penguins became popular symbols of what we would lose to global warming (along with polar bears). Relatively invisible in the public cultural arena, on the other hand, was the growing and unequal affects of the pollution of our atmosphere on marginalized human beings such as Indigenous peoples in the Arctic regions, who are struggling to preserve their culture and societies in the face of rapid climate change. Instead of attention to these issues, penguins have become the newest terrain on which to fight culture wars over human reproduction, while at the same time they have become the latest environmentalist icons. What is the connection between these popular cultural trends? Does it matter in terms of environmental consequences what kind of familial and sexual arrangements we make?

The nature of reproduction

As I have argued elsewhere (Sturgeon, 2009), familial and sexual arrangements are clearly important to environmental issues. Reproduction is an important political issue in our culture, a contested topic in almost every arena of our life. Here, I propose a broader notion of reproduction than customary, using the term "environmental reproductive justice" as a way of connecting environmental issues with social justice issues. In doing so, I am building on the insights of feminists, especially feminists of color and Global South feminists, who have argued for the term "reproductive justice" as opposed to "reproductive rights." "Reproductive justice" refers to more than the mainstream conception of "reproductive rights" (i.e., access to abortion, birth control, the morning-after pill, etc), attempting to address the need to access the means of supporting and nurturing children (i.e., childcare, healthcare, prenatal care, freedom from coerced sterilization, healthy environments, clean air, food, and water, adequate housing, etc), not just allowing individual women to control whether or not they become pregnant (Silliman et al., 2004).

It is important to try to think differently and clearly about these interrelated questions. But how we reproduce — whether we are reproducing people, families, cultures, societies, and/or the planet — is politicized in several layered and contradictory ways. Ironically, given the extreme consequences of certain human models of reproduction for the environment, appeals to the natural are one of the standard ways this politicization of reproduction is obscured.

In short, the politics of gender are about both the politics of reproduction as well as the politics of production — the intertwined ways that people produce more people, manage bringing up children, figure out how to do the work at home at the same time as the work that brings in a paycheck, decide how and where to buy food, clothing, shelter and transportation, take care of elders, and create and maintain all of the social institutions that surround this work. And all of this is central to whether or not our ways of living cause environmental degradation.

Furthermore, these social arrangements are heteronormative: naturalized by assumptions about human relationships — sexual, affective, generational, economic and institutional — that assume as a foundation a particular family form, embedded within a romance plot involving narrow views of male and female attraction, differentiated gender and work roles, and unequal power relations. Yet, we are encouraged to think of these sexual/social arrangements as "only" personal (Berlant and Warner, 1998), a matter of individual choice (in the liberal version) or of natural/divine determination (in the conservative version). To the contrary, such a heteronormative, patriarchal foundation is not just about family and personal relationships, but it structures understandings and consent to matters of citizenship, market relations, nationhood, and foreign policy. Present social and economic structures are based on a tight insistence on the connection between normative heterosexuality (in other words, socially sanctioned, limited versions of only some kinds of heterosexual behaviors, intimacies, and relationships) and "acceptable," "natural," reproduction. The assumption that heterosexuality is the only form of sexuality that is biologically reproductive underlies heterosexism and gives it its persuasive force. Normative heterosexuality is seen as natural and therefore right because it is a form of sexuality that is reproductive. But more closely examined, this logic is not persuasive; sex is not simply about human reproduction in the sense of having babies. After all, given contemporary reproductive technologies and practices, as well as the fact that sexual desire is far more complex and motivated by far more than the potential for pregnancy, actual heterosexual sex is not so closely connected to reproduction as these arguments about its naturalness want us to believe. Otherwise, no heterosexual would have sex unless s/he intended to conceive a child, and no heterosexual would have any kind of sex other than sex that would produce a child. Rather, these heterosexist arguments are usually about preserving and reproducing particular forms of family, social power, and economic practices.

What many other feminist and queer theorists commenting on the use of essentialist ideas of nature to legitimate a conservative form of family values overlook, is that this particular family form, especially when located within a suburban, consumer economy dependent on extremes of global inequality, might be an important origin of our present environmental problems — and that environmental health is an aspect centrally important to reproduction and production. When heteronormative family forms are bound up in environmentally dangerous social and economic practices, we have a situation in which we are promoting environmental damage by naturalizing heteronormative patriarchy, preventing us from imagining and putting in place alternative ways of living more lightly on the earth. Thus, resisting and/or critically evaluating claims to the natural is an essential method of enabling people to consciously create better, more environmentally sound, and more socially just arrangements of work and life.

In this essay, I explore various stories about reproduction found (or conversely, made invisible) in contemporary popular culture, in an attempt to think about how our accepted ideas about the nature of babies, families, marriage, populations, genes, and parenting intertwine with and influence our understanding of environmental issues, or what might be called planetary reproduction, an approach that could be labeled "environmental reproductive justice."

Penguin family values: sexuality in nature

I'll start with competing popular versions of what might be called "penguin family values," that is, the use of the sexual and mating habits of penguins as tokens in the culture wars over the naturalness of heterosexuality or homosexuality. As I mentioned at the start, the Academy Award-winning 2005 nature documentary, *The March of the Penguins*, drew a surprise fan base: right-wing fundamentalist Christian evangelicals. According to the *New York Times*, some conservative religious ministries encouraged their families to attend *The March of the Penguins* together and to write about their spiritual responses according to prompts provided by their pastor. The conservative film critic and radio host Michael Medved was quoted as saying: "[*The March of the Penguins*] passionately affirms traditional norms like monogamy, sacrifice and child rearing...This is the first movie [traditional Christian audiences] have enjoyed since 'The

Passion of the Christ.' This is 'The Passion of the Penguins'" (Miller, 2006).

Particularly odd about this promotion of the penguin family as the ideal Christian family was the equal gender division of labor depicted in the film. Though conservative Christians claim traditional family values involve a complementary appreciation of women's work and men's work, each having a valued and necessary place in the family, the patriarchal framework of the husband acting as Christ to the wife as his domestic helpmate belies true equality. Unlike the idealized patriarchal division of labor that fundamentalist Christians espouse, the division of domestic labor by the penguins is not complementary but rather more strictly equal. After the egg is hatched, the male penguin takes care of it by balancing it on his feet, while the female penguin is the first of the pair to make the arduous 70 mile trek back to the water to get food for the baby chick. When the females return, the males transfer the now-hatched chick to them for care, feeding and warmth while they make their trek to the ocean in turn. Both leave to forage for food and both care for offspring.

One possible way to understand the right-wing Christian fondness for the penguin's arrangement of sharing domestic labor, so unlike what they usually promote, is the effect of the heroic way the males are portrayed, daddies suffering collectively to protect their young against the brutal cold and blinding snowstorms. Clearly, this is how patriarchs should protect their families, with complete commitment and at risk to themselves. The female penguins in the movie, though also sacrificing their health and well-being for their babies, somehow aren't as moving in their long arduous walk as the huddled mass of penguin dads toughing it out together through the arctic night; neither is their equally long wait for the males to return an important part of the narrative. Such a heroic portrayal may also be a way of unconsciously taking out the sting of the material reality that under the conditions of postindustrial global capitalism, women are often co-breadwinners, and men may have to do more domestic labor to keep the family going. Another aspect that might have been attractive to social conservatives might have been the way the film closely connects romance (or desire) with the goal of having children and giving birth, avoiding the messy reality of polymorphous human sexuality. In doing so, *March of the*

Penguins is following a standard anthropomorphic script of TV. nature shows, in which animal mating and reproduction is consistently represented as a metaphor for human heteronormative romance and nuclear families (Wilson, 1992).

What really made this adoption of penguins as promoters of a moral majority so ironic, however, was the already iconic status of penguins as devoted gay couples and parents. The bonding of same-sex penguin pairs, it turns out, is not only fairly common but was actually enjoying an unprecedented amount of publicity in the two years just before the film arrived. In fact, as several letter writers to *The New York Times* pointed out in their response to the article, the disjuncture between these two popularized images of penguins just showed how radically separated from each other are communities of gay people and communities of right-wing religious conservatives: if the Christian fundamentalists had just bothered to google "gay penguins" or even "penguins," they would have immediately encountered a number of gay penguin sites, including the story of Roy and Silo, the Central Park Zoo gay penguin couple about whom a children's book was written and the saga of the gay penguin community at a German zoo. They should have been prepared by the popularity of the penguin as a symbolic saboteur of Christian conservatism; if so, they wouldn't have been so surprised and outraged by the liberal tolerant moral of the children's film *Happy Feet* (2006). It's worth taking a little closer look at each of these cultural phenomena to see how discourses of the natural are flexibly used in the culture wars around sexuality.

Roy and Silo were two penguins that lived at the Central Park Zoo, and who were deeply bonded to one another. As is often the case, because penguin genitalia are not obviously sex-differentiated, the keepers did not know that the pair was same-sex until they noticed that an egg was never produced. Upon closer examination (necessitating, most likely, a DNA test to distinguish sex), the keepers discovered that Silo and Roy were both male. Though the couple went through all the usual courting displays, sexual activity and nest-building behaviors, they were missing an essential element of their reproductive ambitions: an egg. The zookeepers decided to help them out by providing them with another penguin's egg. Roy and Silo successfully raised their egg into a healthy chick, named Tango. This charming penguin family romance was memorialized in the children's book *And Tango Makes Three* (Parnell and Richardson, 2005). The book proposes the moral that all kinds of families, and all kinds of reproductive methods, are equally valuable as long as love, stability and nurturing are involved. As a back-cover blurb of *And Tango Makes Three* by well-known openly gay actor Harvey Fierstein says: "This wonderful story of devotion is heartwarming proof that Mother Nature knows best." The assertion that love and parenting naturally — indeed, biologically — come in both heterosex and same-sex forms was a moral lesson based on nature that enraged many right-wing religious homophobes. Right-wing religious activists in a number of U.S. communities sought to keep *And Tango Makes Three* out of libraries and schools (Huh, 2006). Roy and Silo's story and the publication of the book were not just innocent and diversionary stories, but were cultural tokens of the political contest around gay marriage and gay parental and adoption rights. For human beings, the Central Park penguins were made into a living symbol of the naturalness and success of gay marriage and, depending on one's position in this contest, were celebrated or excoriated by humans as a result.

Of course, *And Tango Makes Three* did not cover the continuing saga of Roy and Silo's relationship. After raising Tango, they eventually broke up, and Silo became sexually involved with a female penguin named Scrappy. Some of New York's gay community took it hard that the apparently committed relationship of Silo and Roy was not as solid as they had hoped. Given the powerful legitimating force in the United States of the idea of nature underlying what is acceptable in human behavior, these cultural contests were about serious, material issues, particularly for gay, lesbian, bisexual, and transgendered people's lives. For queer activists involved in struggling against right-wing attacks on their communities and families, the relationship of the gay penguins served as welcome proof of the natural nature of same-sex love, romance, parenting and domestic stability. In their lives, threatened by heteronormative institutions, which far from protecting their relationships were openly hostile to them, the symbol of the gay penguins was not a trivial thing. Presumably, they were therefore relieved when, in another twist of the story, Silo abandoned the female penguin, Scrappy, and returned to Roy.

Ironically, though both the gay community and the religious right have been invested in the symbolic importance of penguin monogamy and long-term pair bonding, an assumption of the permanence of penguin bonds appears to be problematic in terms of actual penguin behavior. Penguin sexuality, it turns out, is quite variable, with breeding behaviors based on homosexual, heterosexual, trios, quartets, and single parent relationships. Within all species of the genus penguin, partners frequently break up and choose another mate after a season or two of reproductive pairing, though some, such as the Humboldt penguin, frequently form very long-term, multi-year bonds. Penguin family values may include monogamy, but usually only if it is serial, and it doesn't seem to matter too much to the penguins if it is same-sex or hetero-sex monogamy (Bagemihl, 1999).

Arguments from the natural about sexuality, of whatever kind, especially when one uses penguins as one's touchstone, turn out to be pretty slippery (Haraway, 1995; Alaimo, 2010). In general, the sexual practices of animals are so variable that little can be proved about human sexuality using animal examples, though it is a common narrative in popular culture. Furthermore, as Roger Lancaster points out, though there might appear to be short-term advantages to arguing that gayness is biological, inherent, and therefore natural and immutable, there are serious dangers in using these arguments. Not only do arguments from nature about sexuality play into the logic of conservative versions of the family as well as biological determinism, but they carry very dangerous possibilities for many people:

> For gays can only be gay "by nature" in a "nature" that already discloses men and women whose deepest instincts and desires are also different "by nature." In the resulting sexual imaginary, biologically engineered "real" men are always in hot pursuit of "real" women, who always play coy. In such a paradigm, every conventional gender norm, down to the last stereotype, is attributed to a fixed, immutable biology...Norms reified; men and women trapped in their "natures"; a radical division of gay people from straight people, of queer sex from normal sex, of our experiences from theirs...One scarcely has to imagine extreme scenarios to see that this is not good for gay people. Or for straight people, either. (2003: 280)

Finally, as I have stated before, in both pro-heterosexist and pro-gay cases, arguing for the naturalness and superiority of the U.S. nuclear family form ignores its implications in environmental problems. But this does not prevent penguin family values from playing a role in environmentalist popular culture.

Environmentalist penguins fight back

In fact, if one were worried not about what Emperor penguins might symbolize for human sexual mores, but about the penguins' own reproductive health, one would not focus on their domestic, political, or sexual arrangements, but rather on the important relationship between their biology and their environment, their adaptation to their particular environmental niche. Emperor penguins are supremely and exactly suited to the particularities of their challenging Antarctic climate, and the method of protecting their eggs and raising their chicks that so thrills both the Christian right and the gay penguin supporters is the only way that they have managed to maintain their population and continue their reproductive cycles. Whether this adaptation demonstrates an intelligent design or not could no doubt be a point of debate, depending on whether one admires the amazing feat of the survival of the penguins in such a demanding climate, or whether one would want to argue that had the designer been intelligent, s/he would have provided a more secure and sensible warm spot for the penguin's egg other than balancing it on two very hard and wobbly feet. Nevertheless, to see the Emperor penguin as just a survivor is to miss a central part of its existence: that it is matched in specific and fairly inflexible ways to its particular environment. The Emperor penguin is not a survivor but an integral element of its environment, existing nowhere else but the Antarctic. This element of integration with and dependence upon environmental particularities is something we are comfortable with when thinking about animals, but not when we are thinking about human societies, because our dominant frameworks see us as separate from and in control of nature.

What focusing narrowly on mating habits as political signifiers misses is the undeniable fact that the penguin's Antarctic environment is rapidly changing, due to the warming trend called global climate change. The southern polar ice is melting at a rate faster than at any other time in the geological record. For the Emperor penguin, this means a farther and farther walk to find ice thick enough to support the huddled penguins for the length of time needed for the birth and raising of the penguin offspring. This imminent threat to the existence of the Emperor penguins as a species is the unspoken backdrop to *The March of the Penguins*, as the director, Luc Jacquet, is willing to admit. The director and the producers deliberately did not mention global warming in the movie, as they were worried about giving the movie a "political" message (Miller, 2006). But they clearly did hope the movie would raise people's consciousness about the beauty and value of the penguins, so that when discussion of their status as a newly endangered species becomes more well known, people would have sympathy for these special animals and be interested in saving them.

Indeed, the movie does seem to have produced a widespread attachment to the image of the penguin as a symbol of good and beauty, especially when portrayed as under environmental threat. For example, Tom Brokaw's 2006 Discovery Channel documentary, *Global Warming: What You Need to Know*, also uses images of Antarctic penguins to drive home the danger of the melting polar ice. The good feelings produced by portraying penguins as the ultimate in natural family and moral behavior are manipulated into environmental concerns by stressing the endangered status of the penguins. This led to a rise of the use of penguins in advertising of the period. As pointed out in an article in *The New York Times*:

> "There's obviously something about these little guys that is leading advertisers to think it says something about us as consumers to associate ourselves with penguins," said Michael Megalli, a partner at Group 1066, a corporate identity consulting company in New York. One theory Mr. Megalli offered is what he called "the Al Gore thing" — that is, "we want to reassure ourselves penguins will have a place in a world with global warming." (Elliot, 2007)

A 2008 Ad Council promotion for the organization Earth Share shows a family of Emperor penguins (large parent [mother? father?] watching over three smaller penguins) on a grassy expanse, under the caption: "How can you help protect the prairie and the penguin?" as though penguins and prairies could ever be found in the same place. Once again, penguins are removed from their singular habitat and constructed as a human-like family (Emperor penguins only have one chick at a time) in order to appeal to a desire to protect "the prairies and the penguins and the planet."

This use of penguins as symbols of good family morals endangered by human-caused environmental problems appears again in the movie *Happy Feet* (2006). In this Academy Award-winning animated children's movie, an Emperor penguin society is held together by their ability and reverence for singing. Each penguin, in an individualistic twist improbable in a species that has almost no visible differentiation (including, as mentioned above, little obvious sexual difference), must find their own "heartsong." Their heartsong not only defines them as singular beings but also is essential to enabling them to find their one single true love and therefore to breed successfully, another popular cultural version of romanticized monogamous heterosexuality determined by nature. The hero of the film, the boy penguin Mumble, cannot sing, however — though he can dance, an ability that is treated with horror and shame by his parents, peers, and elders. Like homosexuality, his desire to dance not only is different, but it threatens him to a life of infertile relationships with other different outcasts (such as the small band of "Latino" Chinstrap penguins he befriends), because without a heartsong, he will not attract a mate.

When Mumble's difference is blamed for the decreasing availability of fish that is causing a famine for the Emperor penguin society, he is cast out by the high priests and vows to find the cause of the food scarcity, which turns out to be over-harvesting by human fishing corporations. By the end of the movie, Mumble's penguin charm, along with his dancing ability, has mobilized human beings to stop over-fishing. His love of his community, his success in bringing back the fish, and his over-the-top dancing also destroy the intolerance of his community to difference. Not incidentally, Mumble ultimately wins over the (female) love of his life and ends up in a happy, heterosexual, and successfully reproductive nuclear family. That the resolution of all of his problems seems to require the restoration of his "natural" status as the head of a heterosexual family undermines the message of inclusion around sexuality quite a bit, but still didn't prevent Christian conservatives from being outraged at the barely disguised attack of the movie on their positive interpretation of *March of the Penguins* (Medved, 2006).

But even beyond the extremism of certain very conservative Christian commentators, the influence of *The March of the Penguins* and *Happy Feet* have combined to promote penguins as popular symbols that constantly conflate heterosexist family ideals and the need to resist "environmental" threats. Cashing in on the popularity and specific connection made between normality, healthy, happy families and penguins, Roche Pharmaceuticals contracted with the copyright-holders of *Happy Feet* to produce an ad campaign about the use of their products for the flu. A 2006 Roche ad in Newsweek, depicting a mother penguin from *Happy Feet* protecting a baby penguin, uses the caption: "It's Flu Season: Protect Your Family Like Never Before."

There is no question that real, as opposed to symbolic, penguins are in fact endangered by human-caused environmental problems. Penguin reproduction, as individuals and as a species, is closely dependent on systems of planetary reproduction as huge as global climate systems. Presently, those systems are set on a course of rapid warming by the carbon and methane emissions produced by human industrialization, air pollution and factory farming practices, economic practices not unrelated to the high consuming, decentralized formation of the U.S. nuclear family structure. Seeing the penguins as representative of natural human family forms, whether hetero- or homosexual, completely misses the actual nature of the penguin's reproductive system, which is interfused with the Antarctic environment. The lesson of the penguins is not a lesson in intelligent design, or in patriarchal heroics, or in the naturalness of gay marriage; rather it should be a lesson in the ways in which human social reproduction is interrelated with environments both regional and planetary, and vice-versa.

Deconstructing polar opposites: endangered peoples, endangered cultures, endangered natures

It is interesting that in the face of this popular cultural emphasis on the negative environmental effects of climate change on animal reproduction and hence survival, there is little mention of people's reproduction and survival, and where it appears, it often only stresses problematic ideas about particular people's supposed over-reproduction. Missing from the popular culture arena, for the most part, is any attention to the immediate threat to numerous groups of people especially vulnerable to climate change by reason of geography, poverty, or political discrimination. The use of a group of people as symbols of endangered species is uncomfortable for the authors of popular culture (as it should be), partly because it calls up questions of unequal responsibility and unequal consequences that are difficult to deal with in the arena of popular culture as entertainment. Both recent global warming documentaries mentioned above, Gore's and Brokaw's, use images of penguins and polar bears to dramatize the consequences of melting polar ice, but neither mention the impact of climate change on Arctic Indigenous peoples, one of the groups of people already most seriously impacted by climate change. And I put this story together with the story of the Emperor penguins with trepidation, since Indigenous people are not penguins, and endangered tribal cultures are not endangered species; i.e., Indians are not animals. Seeing Indigenous people as endangered species and thus equating them with animals is dangerous because such depictions can be racist. Such a parallel re-enacts the questionable trope of the "disappearing Indian," a dominant narrative that discounts and obscures the struggle of real Indians to exist and successfully transform their cultures strategically for survival. Arctic Indigenous tribes may be threatened by climate change, but they are resilient and experienced in resisting threats to their people. As Chickaloon Grand Chief Gary Harrison, says in the film *Through Arctic Eyes* (2005), a movie documenting the effects of climate change on Arctic Indigenous people: "We've adapted in the past, which is why we are still here."

Yet as a story about the environmental politics of reproduction, the ways in which cultural reproduction needs to be valued as much as biological reproduction, the relation of planetary reproduction to human reproduction, and the need to comprehend human beings as embedded in environmental systems on which they are dependent, the experience of the Indigenous peoples of the Arctic region needs to be more widely known. As many sources note, because of the rigors of survival in Arctic areas, Arctic native peoples are necessarily close to an environment that is particularly sensitive to the effects of climate change, a region tied to the global ecological system in so many intricate ways that changes in the Arctic have world-wide consequences.

Rather than seeing themselves as an endangered species, vulnerable and helpless, Arctic First Nations have been politically active in publicizing the problem of global warming and suggesting solutions for many years before other people paid attention to the issue. They have known that the threats they face to their culture and livelihoods are early warnings for the threats people around the world will face. Patricia Cochran, former chairperson of the Inuit Circumpolar Council, points out the world-wide implications of what Arctic First Peoples are experiencing now:

> All of this will have a profound impact on the viability of Indigenous cultures throughout the North, and further afield. Everything is connected in nature; what happens in Alaska will affect all other places of the world as a cascading effect, as scientists call it, will occur. (Cochran, 2007)

Global warming has had a measurable impact on the lives of Arctic Indigenous people for many years. The rapid melting of the sea ice has severe consequences for marine life, migration of caribou herds, and coastal villages (because sea ice creates a buffer for coastal settlements against large waves). Increased thunderstorms and lightning cause more forest fires. Dangerous levels of UV radiation cause increased incidence of skin cancer and damage to eyesight. Ecological stress and disruption to traditional plant and animal food sources force a turn to a diet of store-bought foods that cause diabetes (Krupnik and Jolly, 2005; *Through Arctic Eyes*, 2005).

Indigenous activists and scientists emphasize that the extensive knowledge they have about sustainable practices, whether it is supported by scientific expertise or traditional experience, is knowledge that arises from a particular way of life, one that needs to be respected and maintained. It is not knowledge gained from the mystical identity of being Ecological Indians (Sturgeon, 2009), but sophisticated information needed now by all those, Indigenous or not, trying to understand and redress climate change. Material practices of reproduction and production have epistemological implications, that is, they affect what we know and how we know it. We are all, like the penguins, suited or not suited to particular ecological contexts and living without respect for those contexts has consequences. Worse, from the perspective of Indigenous activists, global capitalist industrialized ways of living and (not)knowing threaten the existing knowledge base needed to live in sustainable ways not least because the cultural existence of Indigenous people is threatened.

Randel Hansen calls this situation the "ethnocide via climate change of Arctic Indigenous communities" (Hansen, 2005). How do we think about this, as these processes accelerate so that, within a generation, these Indigenous communities may not survive? Is this "natural"? What kind of environmental politics can encompass both the threat to Emperor Penguins and Alaskan Natives from global warming? The disjuncture between the politics of species preservation and the politics of environmental justice presents a barrier to thinking through the relation between these looming disasters. For instance, in Al Gore's 2006 film, *An Inconvenient Truth*, there is a lengthy discussion of the consequences of the melting of the northern and southern polar ice. There is no mention at all of the consequences of this drastic change on Arctic Indigenous peoples. Instead, there is a wrenching depiction of an animated polar bear trying unsuccessfully to get onto a melting ice floe in a vast ice-less sea. Yet the ecologies of the polar bear and of the Arctic Indigenous peoples are interrelated, and surely, both are worth concern and intervention. They are also ecologies interrelated with industrialized ecologies. The reproduction of industrialized economic systems, particular by the United States and Western countries, has consequences for planetary ecological workings on a global scale as well as on the scale of communities, families, and species, determining the ability of animals, families, and cultures to reproduce in healthy and sustainable ways.

Relevant to my earlier discussion of the penguins, we should understand family structure in these Indigenous communities as arising from interrelationships among animals, land, and economic practices. Family does not float free in nuclear groupings of two adults (heterosexual or not) and two children, independent of the consequences of their material practices whether they are industrialized or hunter-gatherers. The idea that families are either separate from or purely reflect a romanticized or anthropomorphized "nature" is an illusion — whether they are Western or Indigenous or any other kind of family. So the point is not that we all have to or should replicate the family structures of Arctic Indigenous peoples (which are various), but we might try, as environmentalists, feminists, and gay activists, to be cognizant of the material interrelationships produced by particular familial forms so that we can choose responsibly ways of living, producing, consuming and reproducing on our planet. Romanticizing Indigenous people, or ignoring the technological and ecological underpinnings of all ways of living, are different forms of racism, both of which can make ecological ethnocide invisible.

Environmental justice family values

Given the issues I have raised with discourses about the use of penguins as symbolizing environmentalist and family values simultaneously, and narratives that make invisible the plight of native peoples, what kind of rhetoric should environmentalists use to bring people to their side?

One thing they probably should *not* do is depict environmentalism as a heteronormative family romance. Such rhetoric obscures the need to put pressure on corporations to change their labor practices — including health care, childcare, pay equity, and global labor practices (all, I would argue, important to real family values) — as part of an environmentalist agenda. In sidestepping these issues, the environmentalist family romance (common also in children's environmentalist films [Sturgeon, 2009]) runs the risk of undermining people's willingness to recognize the ways in which families built on Global North consumerism may need to change their understandings of their relationship with the natural world, and thus their practices of living and working; a critical stance which also requires challenging heterosexist norms. Examples of these family romance plots abound, from both corporate and environmentalist sources.

One environmentalist example is a 2006 ad for the Seaworld/Busch Gardens Conservation Fund, which concludes with the statement: "We all belong to the same family," and, while the caption is replaced by the mission of the Conservation Fund ("Research, Protection, Rehabilitation, Education"), the earth floats beautifully in space, magically free of all the conflict, power, inequality, and policy failure that actually undermine effective environmentalism. The ad closes on a picture of a mountain lake, with the Sea World/Busch Gardens Conservation Fund logo superimposed on it, and the caption: "Conservation: It's in our nature."

Do we really all "belong to the same family"? Is conservation "in our nature"? What does this mean for understanding our environmental problems? What kind of family is environmentally sustainable and how can we encourage such considerations? If corporations like Roche and environmentalists like the Conservation Fund use the same narrative frames and images of nature that legitimate heterosexist white middle class nuclear families, how will we develop a useful critique of that family form's complicity in environmental problems? If the family we think of as natural and normal is white, Western, heterosexual, and middle class, how will we raise consciousness and concern about Indigenous and Global South families, many of which suffer more severely from environmental problems today? The rhetoric of penguin family values limits us to the same idea of what is natural that are promoted by those institutions and corporations that cause environmental destruction. To value nature and to correct social inequalites, we might want to shake off these normative ideas about nature, to see it as more dynamic, more interrelated with human practices, more agentive, and more complicated than we can if we rely just on these dominant stories about nature. We are not outside the earth looking down upon it. We are inside of complex relationships with other biological entities; we impact and are impacted by the interrelationships of those entities. Responsibility to those networks and dynamics can only be brought into view if we understand ourselves as animals among other animals, with varied sexualities, complicated family relationships, and multiple desires — perhaps peculiar animals with astounding abilities, but still part of an interconnected world and thus answerable to it.

* * *

Noël Sturgeon *is Professor of Women's Studies and Graduate Faculty in American Studies at Washington State University. She is the author of* Environmentalism in Popular Culture: Gender, Race, Sexuality and the Politics of the Natural *(University of Arizona Press 2009),* Ecofeminist

Natures: Race, Gender, Feminist Theory and Political Action *(Routledge 1997) and numerous articles on environmental justice cultural studies, antimilitarism, feminist theory, and social movements.*

* A longer version of this chapter originally appeared in Sturgeon (2009).

REFERENCES

Alaimo, Stacey, 2010. "Posthuman desire: Queer animals, science studies, and environmental theory." In Catriona Sandilands and Bruce Erickson, editors, *Queer ecologies: Sex, nature, biopolitics, desire*. Bloomington: Indiana University Press.

Bagemihl, Bruce, 1999. *Biological exuberance: Animal homosexuality and natural diversity*. St. Martin's Press.

Berlant, Lauren and Michael Warner, 1998. "Sex in public," *Critical Inquiry 24* (Winter), 547-566.

Cochran, Patricia, 2007. "Arctic Natives left out in the cold," *BBC News*, Thursday, 4 January.

Elliot, Stuart, 2007. "A procession of penguins arrives on Madison Avenue," *New York Times*, 10 January.

Hansen, Randel, 2005. "The future is now: Arcs of globalization, Indigenous communities, and climate change," Presentation at the American Studies Association Annual Meeting, Washington D. C., November.

Haraway, Donna, 1995. "Otherworldly conversations, terran topics, local terms," 62-92. In Vandana Shiva and Ingunn Moser, editors, *Biopolitics: A feminist and ecological reader on biotechnology*. London: Zed Books.

Huh, Nam Y. 2006. "Schools Chief bans book on penguins," *Boston Globe*, 20 December.

Krupnik, Igor and Dyanna Jolly, eds. 2002. *The earth is faster now: Indigenous observations of arctic environmental change*. Fairbanks, AK: ARCUS (Arctic Research Consortium of the United States).

Lancaster, Roger N. 2003. *The trouble with nature: Sex in science and popular culture*. Berkeley: University of California Press.

Miller, Jonathan, 2006 "March of the conservatives: Penguin film as political fodder," *New York Times*, Tuesday, 13 September, D2.

Parnell, Peter and Justin Richardson, 2005. *And Tango makes three*. New York: Simon and Schuster Children's Publishing.

Silliman, Jael, Marlene Garber Fried, Loretta Ross, and Elena Gutiérrez, 2004. *Undivided rights: Women of color organize for reproductive justice*. Boston: South End Press.

Sturgeon, Noël, 2009. *Environmentalism and popular culture: Gender, race, sexuality and the politics of the natural*. Tucson: University of Arizona.

Wilson, Alexander, 1992. *The culture of nature: North American landscapes from Disney to the Exxon Valdez*. Cambridge, MA: Blackwell.

JELENA VESIC

'Walking on Thin Ice'
The Ice Bear Project, the Inuit and climate change

...we can no longer rely on "experts" who know the "right answers." The alternative is to open decisions to a greater variety of players, disciplines, and voices, and their diversity of values, experiences, and perspectives.
(Lister and Kay, 2000:203)

Introduction

By now it has become a common claim that polar bears are expected to go extinct within the next 100 years (Campbell and Lunau, 2008; Schliebe et al., 2008; Zhu, 2009). Often described as a majestic species of the North on the verge of extinction, pictures, images or video footage show a bear stranded on a chunk of ice in the middle of the ocean. Al Gore's *An Inconvenient Truth* goes even further; it depicts a bear swimming towards a melting ice floe and eventually giving up and drowning due to exhaustion. "They're in trouble, got nowhere else to go," Gore comments, as the image of the struggling polar bear is shown. As Mitchell Taylor, a retired biologist from Nunavut, explains, "Polar bears have become the poster species for doomsday prophets" (quoted in Campbell and Lunau, 2008: 48). Thus, it is no surprise that scientists, environmentalists, animal rights activists and even

Figure 1: A photo of the Copenhagen Ice Bear nearly melted. Once the ice is completely melted and all that is left is the skeleton, the sculpture starts to resemble those of dinosaurs and other prehistoric creatures found only in museums, sending a clear and loud message regarding the future of the polar bear (photo by L. Anders Sandberg).

politicians use the bears as a tool "...in the fight against global warming" (ibid.: 52).

One of the most recent examples of how the species is being utilized to raise awareness and call for action on a global scale is the Ice Bear Project. In partnership with the WWF-World Wide Fund for Nature and Polar Bears International, Mark Coreth, a sculptor and the project founder, along with his team, created a life-size sculpture of a polar bear which was put on display during the Copenhagen Climate Change Summit. As the name suggests, the Ice Bear is made of ice and has a bronze skeleton which becomes only visible as the ice melts, symbolizing the thinning and shrinking of the polar ice caps (Figure 1). Following his trip to the Arctic and witnessing the effect of climate change, Coreth wanted to create an art piece that would "inspire audiences everywhere to make a connection with the Arctic" (IBP, 2010). Described by Coreth as "a message board, a point of discussion... a wake up call," the Ice Bear is meant to "inspire you to explore how you too can help the bear, the Arctic and therefore help bring Nature back into balance" (ibid.).

In Canada, however, polar bears are hunted for both subsistence and trophy-hunting purposes by the Inuit peoples of the North, who now find themselves constantly having to defend and justify the continuation of the hunt. Although the criticism is carefully directed toward trophy-hunting specifically, the fact that Inuit are the ones providing this service to sport hunters makes them a target. A perfect example of this is a statement recently made by Jeff Flocken, the Washington D.C. office director for the International Fund for Animal Welfare; "...before we can address global climate change, we have to try to stop the commercial trade and trophy hunting" (quoted in Zhu, 2009: A2).

This paper interrogates the Ice Bear Project through a climate justice perspective. Climate justice, as articulated by Bond et al. (2007), highlights the disparities in responsibilities for and impacts of climate change, emphasizing the marginalization of peripheral communities (Warry, 2007; Tyrrell, 2006). Such a marginalization may not only be about climate change causing physical hardships, such as melting ice caps and permafrost, droughts and flooding, but also the portrayal of marginalized peoples as victims or villains, and their sources of livelihoods (such as rainforests and polar bears) as a sacred global commons. What remains invisible and obscured by such narratives of victimization, vilification and global appropriation are not only the attempts and struggles of marginalized peoples to preserve their cultures and economies, but also the potential resiliency and ability of them and their ecosystems to change and adapt.

In contrast to the victimization, vilification and appropriation narratives, the situation in the Canadian North shows a strong, viable and respectful relationship between Indigenous peoples and the polar bear. Polar bears are hunted mainly for subsistence purposes; the meat is consumed and the hide is used for clothing and handcraft (Schliebe et al., 2008). Only a small portion of the annual harvest quota (22%) is allocated for sport hunting (Canadian Polar Bear Technical Committee quoted in Dowsley, 2009). But the sports hunt is nevertheless important as Inuit hunters sell their permits to trophy hunters for from $25,000 to $40,000. In 2005, the annual value of 110 sport hunts of polar bears in Canada was $2,090,000 (Dowsley, 2009). The harvesting of polar bears, whether for subsistence or trophy-hunting purposes, provides northern communities with many social, economic and cultural benefits.

The Inuit are skilled and adept in living with and managing the polar bear for their economic and cultural survival (Warry, 2007; Morrison and Wilson, 2004; Brazil and Goudie, 2006; Campbell and Lunau, 2008). Since management was introduced in the 1960s, the world's polar bear population has increased from 5,000 to 20-25,000 bears (Derocher, 2008). Much of the success has been in Canada, where the polar bear population stands at 15,000, hunting quotas have never been exceeded, and the Inuit have been integrated in polar bear co-management (see Dowsley, 2009; Schliebe et al., 2008; Thiemann, 2008). On the last point, Derocher is categorical: "...on the issue of hunting, both (sic) managers, hunters and polar bear scientists are on the same page" (Zarate, 2010).

Even the controversial sports hunt is an integral part of Inuit traditions as well as respectful of the polar bear (Dowsley, 2009). Only Aboriginal peoples with dog teams are allowed to take sport hunters out (Dowsley, 2009), which gives the bear a good chance of escape, by sensing or hearing the dogs — dogs not being as fast as snowmobiles. The use of dog teams also makes the hunt more challenging in unpredictable weather conditions. Furthermore, tags are not transferable if a bear is not killed. These conditions result in only a 50% success rate in the sports hunt, which means that sport hunting quotas are never filled (Campbell and Lunau, 2008). While the polar bear is given considerable agency in the hunt, it may also be able to adopt to climate change by cross-breeding with grizzly bears (Campbell and Lunau, 2008; CBC News, 2010;).

Aboriginal people agree with scientists that climate change poses a threat to polar bears. But while there are declines in some parts of the Arctic, there are healthy populations elsewhere (Schliebe et al., 2008). And if numbers start declining, the Inuit insist, they will be the first to stop harvesting the animal (Campbell and Lunau, 2008). As one Inuvialuit hunter, says, "If something goes wrong here, we'll know. We live it" (quoted in Campbell and Lunau, 2008: 46). The livelihood of the northern communities depends on animals and therefore, preventing and avoiding over-harvesting is essential.

Given these conditions, it is perhaps not strange that many believe that the management and conservation of polar bears is one of the most successful strategies in the world and should be applied for management of natural resources elsewhere (Freeman and Wenzel, 2005; Tyrrell, 2006; Aaras, Lunn and Derocher, 2006).

Interrogating the Ice Bear Project

The dominant narrative of the Ice Bear Project is "on thin ice" in its portrayal of the polar bear situation in the North. It seldom represents the social, cultural and even economic presence of the Inuit in the North. It also continues to place the polar bear in a precarious situation, making the species synonymous with imminent extinction. The Ice Bear Project, therefore, is a depiction of the dominant narrative on climate change which marginalizes both local communities and their voices. According to Mark Coreth, the Ice Bear creator, "the polar bear is such a wonderful way...of [exploring the Arctic environment]" (quoted in Sharp, 2009). "This isn't just a sculpture of a polar bear," Coreth continues, "it's a sculpture of an entire environment" (IBP, 2010). For Coreth, the polar bear is in trouble. On 26 November 2009, Coreth (2010a) writes: "I find it distressing to hear of the bears pacing the shores waiting to get to ice and food..." In these statements, the polar bear is prioritized, overshadowing the Inuit and their culture, making it easier for people living in Western societies to sympathize with polar bears rather than with the people living in the Arctic regions.

In May 2009, Coreth traveled to northern Canada to study the polar bear. He then visited a remote Inuit community and spoke to local hunters. On 15 May 2009, he wrote:

We met up with some Inuit hunters... hunting for seal and POLAR BEAR... they just missed one that dived into the sea from the floe edge and escaped... bears do not like men! I have got to find a bear before the hunter does. I spoke to Abraham and Patrick (our guides) about the traditions of bear hunting and about the respects between bear and man. I spoke also to them about the frustrations that they feel with the southerners views of the Arctic and its wild life, about the bears and the uses of all body parts. Kabloona (Southerners) have to respect the Inuit traditions. (Coreth, 2010b)

While Coreth seems to recognize the Inuit perspective, even acknowledging the sacred relationship between the Inuit and the bear, he does not explore it further. And although he is also clearly aware of the Inuit's frustrations with how southerners perceive the Arctic, Coreth does not follow up on it. Instead, he creates a divide between bears and people by pointing out that bears do not like men. This is in contrast to the Inuit, who see the bear as offering itself up during the hunt (Morrison and Wilson, 2004). The dichotomy between people and nature in

Coreth's account is characteristic of a dominant discourse which positions people outside nature (Cronon, 1983). In addition, Coreth successfully distances himself from the hunters, thereby creating a tension between himself and the Inuit.

The narrative surrounding the Ice Bear Project never truly acknowledges the importance and significance of the hunt to Inuit culture. This is reflected in the policy of Polar Sea Adventures, the ecotourism company that arranged Coreth's trip to the Arctic. It claims to value "the enduring Inuit hunting culture" (PSA, 2009). However, the company promotes adventure tourism "for Inuit young and old to showcase their culture to people from all over the world..." (ibid.). The Inuit are here transformed from hunters, who interact with the bear viscerally through the use of all its parts, to guides who merely track and show off the bear to visitors. Such a transition from hunters to guides repeats a transformation that took place in the south more than a century ago. Besides earning a significantly lower income working as guides, there are many problems associated with this line of work, including the loss of cultural heritage (see Dowsley, 2009). What is most troublesome is that the transition, which is a by-product of the dominant discourse, is forced, resulting in disempowerment of local communities while reinforcing the "current unequal power structure" (Bond et al., 2007: 122; Warry, 2007; Isla, 2005; Escobar, 1998). Consequently, Inuit and their culture become nothing more than, as Taiaiake Alfred (2005:41) argues, "a tourist attraction" for elite, privileged westerners.

Another problematic aspect of the Ice Bear Project is that it draws exclusively yet selectively on modern science. Coreth consults Andrew Derocher, one of the leading Canadian scientists on the polar bear to get his information. On 5 October 2009, he writes that in order to make informative and educational decisions it is necessary to "go and speak to those in the know, the scientists who are dealing with the polar bear and its environment"(Coreth, 2010a). There is no mention of the traditional and practical ecological knowledge that the Inuit possess. The references to Derocher are problematic in other respects. Derocher only speaks about disappearing ice caps and the plight of the polar bear, though he elsewhere acknowledges the relationship between the Inuit and the polar bears, the significance of the Inuit hunt, and his support for the hunt (Schliebe et al., 2008;

Derocher, 2007; Derocher et al., 2004; Derocher quoted in Zarate, 2010).

Escobar (1998) and others argue that the loss of natural resources is often seen strictly as a scientific and global problem to be solved by experts (Tyrrell, 2006; Isla, 2005). In the case of the polar bear, hunting is being highlighted as a problem, while the role of the carbon economy is downplayed. Thus, those with the least political power and least responsibility for climate change are blamed for what is happening with the species' population (Escobar, 1998; Isla, 2005; Anand, 2003).

Coreth nevertheless identifies the key role that government and industry play in climate change in the fate of the polar bear. He does, however, place an equal burden on the individual, be it the Inuit hunter or the southern consumer. He writes on 30 November 2009:

> From my perspective having spoken to numerous scientists now... the Ice Bear is correct to the science and has a real purpose in asking people to discuss how we all can make that difference. Government and industry have to tackle the large percentage but in the end we all have to take on the rest... and ultimately the responsibility. (Coreth 2010a)

The individualization of responsibility, as Maniates (2002) refers to it, allows allocation of accountability from the government and industry to the individual. This adversely affects the Inuit whose rights to the polar bear are jeopardized. It also brings guilt and blame to the individual consumer for the plight of the polar bear, a condition exploited by Coreth and environmental groups, while deflecting responsibility from the big industrial carbon producers.

Conclusion

Informed by science, and fueled by media coverage, the Ice Bear Project confirms a dominant narrative of climate change which paints a declensionist, pessimistic and dystopic story. Climate change is considered a global problem calling for global solutions. The polar bear is seen as an international problem, though only 3% of Canada's population, the Inuit, inter-acts intimately with the species (Freeman and Wenzel, 2005). The melting Ice

Bear at Copenhagen's Climate Change Summit illustrates and contributes to the dominant discourse by furthering the polar bear as a global icon of climate change which belongs to a global commons.

A climate justice discourse provides an alternative to the dominant discourse. It emphasizes the Inuit's right to the polar bear hunt and their right to use hunting quotas as they see fit, even if it entails selling a portion of them to trophy hunters. The focus is shifted away from the polar bear to the Arctic people and their environment, which polar bears are part of. In simple terms, a climate justice narrative calls for, as Anand (2003) explains, an incorporation of justice and fairness to the debate on climate change. This is important because "an environmentally healthy planet [is] impossible in a world that contain[s] significant inequalities" (Miller quoted in Anand, 2003: 56).

As Dowsley (2009: 171) explains, "[the] symbol of the dying (often drowning) polar bear that is driving international concerns about the species' conservation" is highly problematic, especially in terms of its impact on local Inuit communities. The solution to this problem is complex but must include one aspect: listening to the Inuit.

* * *

Jelena Vesic is a student in the Master of Environmental Studies program at York University. Her main research interest focuses on political ecology and its application to the contestations and negotiations surrounding wildlife conservation and environmental management in Canada.

REFERENCES

Alfred, Taiaiake, 2005. *Wasase: Indigenous Pathways of Action and Freedom*. Peterborough, ON: Broadview Press.

Anand, R., 2003. *International Environmental Justice: A North-South Dimension*. "Climate Change," pp. 21-60. Burlington, VT: Ashgate.

Bond, P., Dada, R., & Erion, G., 2007. *Climate Change, Carbon Trading and Civil Society: Negative Returns on South African Investments*. South Africa: University of KwaZula-Natal Press.

Brazil, Joe and Goudie, Jim, 2006. A 5 Year Management Plan (2006-2022) for the Polar Bear/Nanuk (*Ursus maritimus*) in Newfoundland and Labrador. Wildlife Division, Department of Environment and Conservation. Government of Newfoundland and Labrador and the Department of Lands and Natural Resources, Nunatsiavut Government. 25pp.

Campbell, C. and Lunau, K., 2008. "The War Over the Polar Bear," *Maclean's*, 121, 4-5, 46-52.

CBC News, 2010. "Bear Shot in N.W.T was Grizzly-Polar Hybrid," *CBC News*, 30 April. http://www.cbc.ca/canada/north/story/2010/04/30/nwt-gro-lar-bear.html, accessed 28 June 2010.

Coreth, Mark, 2010a. Blog. *The Ice Bear Project.* http://www.icebearproject.org/blog/, accessed 5 June 2010.

Coreth, Mark, 2010b. The Expedition – Mark's Diary. *The Ice Bear Project.* http://www.icebearproject.org/expedition.html, accessed 5 June 2010.

Cronon, William, 1983. *Changes in the Land: Indians, Colonists, and the Ecology of New England*. New York: Hill and Wang.

Derocher, A.E., Lunn, N.J., & Stirling, I., 2004. "Polar Bears in Warming Climate," *Integrative and Comparative Biology*, 44, 163-176.

Derocher, Andrew (guest), 14 April 2007. Polar Bears in Canada's Arctic [radio show]. In David Fisher (producer) *The Science Show*. Australia: ABC Radio National. http://www.abc.net.au/rn/scienceshow/stories/2007/1893380.htm, accessed 13 June 2010.

Dowsley, Martha, 2009. "Inuit-Organized Polar Bear Sport Hunting in Nunavut Territory, Canada," *Journal of Ecotourism*, 8, 2, 161-175.

Escobar, A., 1998. "Whose Knowledge, Whose Nature? Biodiversity, conservation and the Political Ecology of Social Movements," *Journal of Political Ecology*, 5, 53-82.

Freeman, M.M.R. & Wenzel, G.W., 2006. The Nature and Significance of Polar Bear Conservation Hunting in the Canadian Arctic. *Arctic*, 59, 1, 21-30.

IBP [Ice Bear Project], 2010. "Home" and "Ice Bear Info". http://www.icebearproject.org/, accessed 3 June 2010.

Isla, A., 2005. "Conservation as Enclosure: an Ecofeminist Perspective on Sustainable Development in Costa Rica," *Capitalism, Nature, Socialism*, 16, 3, 49-61.

Lister, Nina-Marie E. and Kay, James J., 2000. "Celebrating Diversity: Adaptive Planning and Biodiversity Conservation," pp. 189-218. In Bocking, Stephen, editor, *Biodiversity in Canada: Ecology, Ideas and Action*. Peterborough, ON: Broadview Press.

Maniates, M., 2002. "Individualization: Plant a Tree, Buy a Bike, Save the World?" pp. 43-66. In Princen, T., M. Maniates & K. Conca, editors, *Confronting Consumption*. Cambridge, MA: MIT Press.

PSA [Polar Sea Adventure], 2009. http://www.polarseaadventures.com/index.htm, accessed 15 June 2010.

Schliebe, S., Wiig, Ø., Derocher, A. & Lunn, N., 2008. "Ursus maritimus." In IUCN Red List of Threatened Species. Version 2010.1. www.iucnredlist.org, accessed 14 June 2010.

Sharp, Rob, 2009. "Climate Change: the Ice Bear Cometh," *The Independent*, 2 December. http://www.independent.co.uk/environment/climate-change/climate-change-the-ice-bear-cometh-1832170.html, accessed 17 June 2010.

Thiemann, W. Gregory, 2008. "Polar Bear Conservation in a Changing Arctic Environment," 6 February, Guest Lecture, York University, Toronto.

Tyrrell, Martina, 2006. "More Bears, Less Bears: Inuit and Scientific Perception of Polar Bear Populations on the West Coast of Hudson Bay," *Études/Inuit/Studies*, 30, 2, 191-208.

Warry, Wayne, 2007. *Ending Denial*. Peterborough, ON: Broadview Press.

Zarate, Gabriel, 2010. "Scientist Support Nunavut Polar Bear Hunters," *Nunatsiaq Online*, 22 January. http://www.nunatsiaqonline.ca/stories/article/7567_scientists_support_nunavut_polar_bear_hunters/, accessed 20 June 2010.

Zhu, Helena, 2009. "Polar Bears Drop off on Thinning Ice and Bullets," *The Epoch Times*, 10-16 September, A1-A2.

ISAAC 'ASUME' OSUOKA

Operation Climate Change
Between community resource
control and carbon capitalism in
the Niger Delta

The British Petroleum (BP) Deepwater Horizon drilling rig
disaster in the Gulf of Mexico in 2010 demonstrated the dangers associated with the fossil fuels-based energy system. It also
exposed the flaw in giving too much power to corporations to
deal with environmental catastrophes. For months, BP's experts
in the 'developed' United States could not stop the spill. The lim-
itations in the responses of the corporation and the United
States government were easy to notice because of the magni-
tude of the 'accident' that caused the spill. But the often silent
and continuous pollution of communities by petrochemical
plants onshore in the state of Louisiana has not attracted the
same amount of attention. The 'fence line' communities of
'Cancer Alley' share very similar experiences to the 'host' communities of resource extraction and related corporate practices
in many parts of the 'third world'. For the inhabitants of the
Niger Delta in Nigeria, for example, blowouts and oil spills have
been a routine part of their lives for decades. Petroleum
exploitation has devastated the natural environment and
resulted in impoverishment and social crisis. Since the late
1950s, transnational oil corporations have drilled crude oil for
the export market, and flared the natural gas containing

methane, carbon dioxide and other substances that pollute the local people and land.

As the global concern over climate change has grown, a globally dominant neoliberal discourse based on markets and new technologies has come to be seen as the most efficient system to reduce national and global greenhouse gas emissions. In the 1990s, the corporations responsible for producing fossil fuels were offering alternatives in the form of clean coal and natural gas — considered to produce reduced emissions. By the end of the 1990s, BP had rebranded itself 'Beyond Petroleum'. The company introduced a new logo to conjure up an image of greenery and announced, along with its competitors, plans to invest in alternative biofuels, solar and wind energy technologies. There are also growing efforts to commodify the atmosphere "through new practices of calculation and distribution", as part of a growing attempt at privatising the climate (Lohmann 2009: 25). Through carbon trading schemes, the corporations are set to make additional profits by claiming to reduce emissions, for example, by reducing the amount of gas flaring in the Niger Delta.

This chapter presents the argument that it is not possible to tell the story of fossil fuels in isolation of the history of industrial society, the nation state, and the corporation. Within the different nation-states that currently claim sovereignty over resources and peoples, we find the existence of different types and levels of 'joint ventures' which determine how exploitation is organised, who benefits from it, and who is victimised and burdened by it. Under corporate rule, the quest for profits by corporations is the overriding determinant of political and social life, and the character of human relationship with nature. More than ever, there is the quest to territorialise and commodify nature and peoples.

The neoliberalism which carbon exploitation and carbon trading represent is being challenged by peoples and movements around the world. I will present a story of how young people in the Niger Delta of Nigeria launched Operation Climate Change in 1998 as a programme of mass actions seeking a halt to petroleum exploitation in the region. Through this story, we discover that the exploitation and use of fossil fuels does not only result in climate change. The extraction of oil by the 'joint venture' of state and transnational corporations reinforces the colonial power relations and the dispossession of communities. The fixation on technological and market driven solutions fails to address the demands of communities for 'resource control'.

'Crude' technology, domination and resistance

As early expressions of corporate rule, European corporations and their armed agents literarily occupied territories in Africa, Asia and the Americas. Some countries were originally created and administered directly by corporations with the backing of their imperial home governments. The corporate driven colonial governments set up structures to facilitate exploitation of minerals and other raw materials. With oil replacing coal as the pre-eminent fuel of the emerging capitalist modernity, new oil and gas exploitation resulted in joint ventures in which the state facilitated corporate access and control over communal land.

In the countries of the Gulf of Guinea, for example, crude oil and natural gas were exploited for the export markets in Europe and North America, following a pattern dating back to colonial coal mines (as in Nigeria). In all the countries in the region, production is almost exclusively determined by transnational oil corporations. In the process of exploitation, the natural environments have been devastated and communities impoverished and repressed.

Nigeria was created by the British Royal Niger Company. While the country was still under British colonial rule, Shell commenced crude oil production in 1956. Since then other transnational oil corporations with origins in Europe, North America and Asia have joined the rush for oil in the Niger Delta. By the 1970s, the corporations were operating in 'joint venture' with the state. Under this arrangement, the state-owned oil company, NNPC, holds some equity in the joint ventures while the transnational corporation remains the operator.[1] State legislation, established under military rule, grants government the power to take over farmland and forest from communities if oil is discovered on such land. Meanwhile, the government neglects the oil spills and general destruction of fragile ecosystems resulting from the use of the crude technology by the joint venture operators. There is now a general agreement that the Niger Delta oilfields have recorded the worst continuous incidences of oil spills globally, with aged pipelines and flow-lines easily given way to corrosion and leaks. With parts of the oil-bearing Niger Delta composed of wetlands or seasonally flooded forests and farmland, reckless engineering results in loss of species. Plants and other organisms are either starved of water or become inundated in water, as oil industry infrastructure disrupt the hydrological regime through construction of access roads and pipelines that lead from wellheads to processing facilities and coastal export terminals.

The crude oil fields of the Niger Delta of Nigeria also contain large amounts of 'associated' natural gas. Since Shell and the colonial authorities determined that there was only a profitable European market for crude oil, the oil infrastructure was built to collect just the crude. The 'associated' gas is burned off at flow-stations upon separation from the crude and water. This practice is referred to as gas flaring. The waste water and other drilling waste are also deposited in the immediate environment. The gas that is flared could have been used to generate energy for Nigerian communities. But Shell and the colonial government did not see any immediate profit potential in this, a situation that continues today.

Figure 1: 'Associated' gas flaring is a source of local pollution (photo by Celestine Akpobari).

'Associated' gas contains significant amounts of carbon dioxide and methane, which are among the major greenhouse gases responsible for climate change.

Paradoxically, however, the deliberate neglect of the corporations to harness the natural gas in the past now provide them with a chance to further increase their profits. In the late 1990s the Clean Development Mechanism (CDM) under the Kyoto agreement offered an opportunity for oil corporations active in Nigeria to claim carbon credits by presenting claims of how new infrastructure would reduce gas flaring.[2] Without any possibility of verifying the claims in the long term, the companies stand to make millions of dollars through paperwork that show how it may reduce pollution that it created in the first place. An independent assessment of the claims for carbon credits through so called gas flare reduction projects revealed that the companies seeking them have been increasing the volume of gas flaring and associated emissions at the same time that they can make profits by selling carbon credits. Meanwhile the communities that have suffered the impacts of oil exploitation (and the colonial violence that preceded and initiated it) get nothing (Osuoka 2009).

In 1998 this author was among the thousands of young people from the communities of the Niger Delta of Nigeria that initiated a series of mass protests and direct action which we called Operation Climate Change. This was a continuation of a new phase of community resistance that began in the late 1980s when Nigerian communities took action to halt the destruction by oil and gas exploitation of their environment and livelihoods. By 1990, the Movement for the Survival of the Ogoni People (MOSOP) proclaimed the Ogoni Bill of Rights, which contained demands of the people for environmental justice as part of a broader demand for political autonomy within a restructured Nigerian state. The MOSOP of the 1990s was an umbrella of organisations that represented women, students and youth, traditional and cultural leaders, and faith based groups. The proclamation of the Ogoni Bill of Rights was followed by mass protests and the shutting down of Shell-operated oil wells. In response to the struggles of the unarmed Ogoni, Nigerian military regimes, with the prompting of Shell, unleashed a reign of terror against the Ogoni. By 1995, the bulk of the leadership of the MOSOP had been murdered by the state.

In December 1998, Operation Climate Change followed the proclamation of the Kaiama Declaration by young people from communities that identify themselves as Ijaw. With the Kaiama Declaration, the Ijaw declared that they "cease to recognise all undemocratic decrees that rob our peoples/communities of the right to ownership and control of our lives and resources, which were enacted without our participation and consent."[3] In this case, the Ijaw followed the Ogoni in demanding restoration of

Figure 2: In 26 May 2009, the Ogoni protest in Bori against Shell and the Nigerian government. The Ogoni have continued to protest the killing of MOSOP leaders, including Ken Saro-Wiwa, in 10 November 1995 (photo by Celestine Akpobari).

community control of land as part of demands for political autonomy for ethnic nationalities. Operation Climate Change commenced as a programme of mass actions to actualise the Kaiama Declaration. It involved *Ogele* processions in towns and villages with singing drumming and dancing. It also involved occupations of oil terminals where people demanded the shutting down of oil wells and the gas flares. The height of Operation Climate Change was the *Ogele* processions of 30 December 1998 which are now remembered for the violent response from the state. In village after village, soldiers deployed by the state opened fire on unarmed citizens. At Kaiama, Mbiama, and Yenagoa people were killed in the streets and women and young girls were raped in their homes as the state unleashed mayhem, ostensibly to defend oil installations.

In 1998, the name given to the mass mobilisation in the Niger Delta was an acknowledgement of petroleum as contributor to global climate change. And the direct actions aimed to shut down oil production and gas flaring was a local response to addressing

the problem. But Operation Climate Change did not only pro-
ceed with the understanding that actions to reduce CO2 emissions was necessary, it also sought to reduce the land grabs for
construction of gas pipelines and Liquefied Natural Gas (LNG)
facilities that are built in Nigeria, Cameroon, Equatorial Guinea
and Angola. Natural gas infrastructure produces very similar
social impacts as crude oil infrastructure. In the countries of the
Gulf of Guinea, both crude oil and natural gas involve the joint
venture of state and corporation and the dispossessing of communities, as production is organised to satisfy energy needs of
distant places.

In the Niger Delta, the factors of state repression, oil industry
security contracting practices, and a thriving black market for
stolen crude oil combined with frustration within communities
to exacerbate violence in the years following Operation Climate
Change. State militarisation of the area was followed by an
armed insurgency in parts of the delta. The mainstream media
increased attention on the area, particularly as militias sabo-
taged oil infrastructure, disrupted production and contributed to
the rise in global crude oil prices between 2006 and 2009. It was
at the height of the armed insurgency in the Niger Delta that the
U.S. Bush administration set up the U.S. Africa Command (US-
AFRICOM) armed forces. Clearly, the United States government
had a view on protecting oil business and supplies from the
Niger Delta and other parts of the Gulf of Guinea where U.S. corporations operate significant fields.

Climate change and carbon capitalism
The neoliberal regime has affected the Nigerian and other
Third World countries' carbon economies in two ways. First,
there has been a tendency for states to promote military
responses against those who challenge the corporate petro-
state. Second, environmental regulations have increasingly
become subordinated to the market. The new system of corporate rule has pushed government agencies to the sidelines. By
the late 1990s institutions like the World Bank and the United
States Agency for International Development were actively supporting reforms that ensured that many Third World Countries
weakened environmental regulations and enforcement in the
extractive sectors of the economy.

With respect to climate change, the period from the late 1980s witnessed the convergence of corporate driven technology and the market in creating often voluntary based carbon 'cap and trade' mechanisms. The efforts to reduce carbon emissions are highly compromised by corporate control and manipulation. For corporations like BP carbon trading and carbon offset projects offer opportunities to continue to increase emissions in their core business of oil production while buying into dubious projects that have the "potential" to reduce or absorb emissions. In reality, carbon trading simply becomes a way for corporations to escape reducing their own emissions. Examinations of particular carbon offset projects within the context of the CDM (Bohm and Dabhi, 2009) and the Reduce Emissions from Deforestation and Degradation program (REDD)[4] (Indigenous Environmental Network, 2009) have shown that in most cases emissions have increased, while corporations make extra profits by selling the offset credits that are obtained by presenting claims of 'hypothetical' emissions reduction.

With schemes like the CDM and REDD gaining ground, there is a new spectrum of blame as communities may now be held responsible for opposing carbon market schemes. If so called forest conservation schemes to absorb carbon emissions fail or are opposed, corporate capital can point to impoverished or "overpopulated" communities as environmental villains. We thus see not just a continuing pattern of marginalisation, but a criminalisation of victims as well.

Proponents of REDD may make reference to capital flows from the developed to underdeveloped countries. The UN-REDD, is already predicting that up to U.S. $30 billion every year could be transferred to underdeveloped countries as payments for REDD projects.[5] But the experience of present and past payments in the form of development aid only masks the reality of overwhelming resource outflows from the poorer countries to the richer. When climate change mitigation schemes are presented in ways that reinforce patterns of national and global dispossession and inequalities, they become part of the problem.

It would be a mistake, though, to conclude that conditions are much different in the developed world. The BP oil blowout from the Deepwater Horizon rig clearly exposes the 'lawlessness' that characterises environmental regulation in the United States.

The U.S. regulatory Minerals Management Service (MMS), Dickinson observes, has "for years ... essentially allowed the oil industry to self-regulate" (Dickinson, 2010). As in countries like Nigeria, Equatorial Guinea and Cameroon, the United States' MMS just serves "to rubber-stamp dangerous drilling operations by BP — a firm with the worst safety record of any oil company — with virtually no environmental safeguards, using industry-friendly regulations drafted during the Bush years" (ibid). MMS staffers, Dickinson continues, are "both literally and figuratively in bed with the oil industry" (ibid).

The shock from the BP disaster followed the shocks from the financial and economic crisis which was experienced in the United States and Western Europe from 2007, with reverbera-tions globally. The big banks and other financial corporations col-lapsed or were at the verge of collapse following failures of spec-ulative investments that proliferated at a time that government and general public sector regulation of the corporations were at their weakest. From both cases we can determine that when cor-porations and the market are allowed to self-regulate there is bound to be disaster.

Conclusion

In the 1990s, the organisations involved with Operation Climate Change in Nigeria understood that crude oil exploitation and accompanying greenhouse emissions through gas flaring cannot be addressed without dismantling the social structures, dicta-torships, state and corporate land takeovers and general abuses of rights that accompany or are produced by industry. The ini-tiators of the Ogoni Bill of Rights and the Kaiama Declaration, therefore, demanded community 'self-determination' and 'resource control', which express their commitment to secure freedoms and spaces for socio-cultural expression and rule. Such demands cannot be addressed by corporations acting voluntari-ly within a market paradigm.

Carbon emissions mitigation via 'carbon trade' reinforces exist-ing patterns of domination and profiteering. Under existing schemes, major emitting northern corporations are allowed to 'off-set' their emissions by securing 'carbon credits' when they make claims for investing in new technologies and projects that may reduce or absorb emissions or absorb carbon dioxide in countries of the South. All of a sudden, the global climate crisis has been seized as a business opportunity with corporations and their backers attempting to bring under their direct control those natural spaces of reproduction of life which have been free for all humanity, and upon which many communities depend for survival.

With corporations like Shell and BP involved in the development of solar, wind and other renewable technologies (and the policies guiding their deployment), there is little or no guarantee that these 'sustainable' technologies would not be deployed in such ways as to entrench the privileging of corporate power over communities. Corporations exist and are structured for profit making and not to address environmental and social problems. Any attempt to seek a positive response from corporations, acting voluntarily, in the form of 'community development' projects or 'corporate social responsibility' only place communities into a corporate stranglehold.

Operation Climate Change acknowledges that the shutting down of gas flares and the deployment of new energy technologies are necessary to mitigate impacts of local pollution and global warming. However, the Ogoni Bill of Rights and Kaiama Declaration that defined the mass actions exposed technological and investment shifts as not sufficient to address the political and economic demands of the historical victims of resource exploitation. For those demands to be addressed, climate change mitigation should be accompanied by a restructuring of power to restore a reasonable degree of sovereignty to communities. Individual enterprises should operate under rules set by a restructured state — a state that protects the freedoms and rights of communities to ample spaces for self-rule.

When corporations are allowed to develop technologies that favour economic growth, there should be no surprise that the environmental concerns figure poorly in their calculations. In the case of BP's offshore crude oil drilling in the Gulf of Mexico, the company's documentation disregarded the possibility of a blowout, spill or an adverse impact on the marine environment and the life contained and dependent on such an environment. By the time the blowout occurred, the government of the United States was clueless on how to respond, waiting instead for BP to launch experiment after experiment as a beleaguered world watched in shock.

Rather than looking up to the BPs of the world, there is a need for changes that involve a decentralisation of energy and technology choices, and a respect for the sovereignty of communities over their energy resources. As Oilwatch, the South-South network of people's organisations has put it, "climate solutions should be based on human security, rectification of ecological debt, land rights, the change of production and consumption patterns, to realize social justice and people's sovereignty. These principles ensure that in the heart of climate solutions are the welfare and interest of the people and the environment."[6]

* * *

Isaac 'Asume' Osuoka is a PhD candidate in the Faculty of Environmental Studies at York University. He continues to work with communities and organisations in Nigeria and other countries in the Gulf of Guinea region to promote change in practices of civil society, governments, oil corporations and international finance institutions.

ENDNOTES

[1] Apart from the NNPC, Nigeria owned oil companies account for less than 5% of production.

[2] The CDM is one of the 'flexible' mechanisms introduced by by the Kyoto Protocol. By means of the CDM, corporations and governments in the high greenhouse emitting developed countries can invest in supposed energy efficient technology and other 'sustainable' ventures in low emitting Third World countries. By this arrangement, it is considered that by reducing or preventing an increase of emissions in the poorer countries, corporations and Northern governments would 'offset' the emissions in the developed countries. The Kyoto protocol unleashed a scramble as corporations devised thousands of schemes to secure additional profits from 'carbon trading'.

[3] "Kaiama Declaration: Communique issued at the end of the first All Ijaw Youth Conference." The conference was held in the town of Kaiama, Bayelsa State, Nigeria on 11 December 1998. For the full declaration, see: http://www.dawodu.net/kaiama.htm.

[4] REDD is an initiative introduced at the eleventh Conference of Parties of the UNFCCC in Montreal in 2005 (Global Canopy Programme 2008). Within the REDD framework, a monetary value is attached to forests based on estimated capacity to act as 'sinks' for CO_2. Based on such estimates, states and other holders of forest titles would be paid to ensure

that forests are not destroyed. While this sounds very much like a great idea at first glance — afterall who will not be pleased with the idea of forest protection — REDD and the similar forest initiatives within the CDM are already encouraging corporate takeovers. With climate change business causing a recalculation and commoditisation of tropical forests, forest dependent communities are at the risk of evictions, as they could now be seen as impediments to maintaining the carbon marketing potentials of forests. There are already reports of 'land grabs' with corporations as well as NGOs seeking legal title. This is reminiscent of how state and oil industry relationships determine how community resources are appropriated.

[5] http://www.un-redd.org/AboutREDD/tabid/582/language/en-US/Default.aspx.

[6] http://www.oilwatch.org/index.php?option=com_content&task=view&id=616&Itemid=48&lang=en.

REFERENCES

Bohm, Steffen and Dabhi, Siddartha, editors, 2009. *Upsetting the Offset: The Political Economy of Carbon Markets*. London: MayFly Books.

Brand, Ulrich and Sekler, Nicola, 2009. "Postneoliberalism: catch-all word or valuable analytical and political concept?" *Development Dialogue*, 51, 5-13.

Dickinson, Tim, 2010. "The Spill, the Scandal and the President," *Rolling Stone*, 24 June. http://www.rollingstone.com/politics/news/17390/111965?RS_show_page=0, accessed 25 June 2010.

Global Canopy Programme, 2008. *The Little REDD Book*. http://www.amazonconservation.org/pdf/redd_the_little_redd_book_dec_08.pdf, accessed 17 June 2010.

Indigenous Environmental Network, 2009. http://www.ienearth.org/REDD/redd.pdf, accessed 8 June 2010.

Lohmann, Larry, 2009. "Regulation and Corruption in the Carbon Offsets Markets," pp. 175-191. In Bohm, Steffen and Dabhi, Siddartha, editors, *Upsetting the Offset: The Political Economy of Carbon Markets*. London: MayFly Books.

Oilwatch, 2010. www.oilwatch.org, accessed 18 May 2010.

Osuoka, Isaac "Paying the Polluter? The Relegation of Local Community Concerns in 'Carbon Credit' Proposals of Oil Corporations in Nigeria," pp. 86-97. In Bohm, Steffen and Dabhi, Siddartha, editors, *Upsetting the Offset: The Political Economy of Carbon Markets*. London: MayFly Books.

UN-REDD, 2010. http://www.un-redd.org/, accessed 15 May 2010.

TANYA GULLIVER

Broken Pieces, Shattered Lives
The lasting legacy of Hurricane Katrina

Shortly after I began teaching a course on homelessness at Ryerson University, in Toronto, Hurricane Katrina hit the New Orleans area, leaving more than half a million people homeless. After a close friend travelled to New Orleans in February 2006 to help Common Ground — a local grassroots organization — with the efforts of rebuilding, I invited her to my class to talk about her experiences. Many students begged me for the opportunity to do the same. After much discussion and planning, Ryerson agreed to let me develop and teach a field course in New Orleans. I have since embarked on doctoral studies that address issues of post-disaster recovery using Katrina as a case study, and I have travelled regularly to New Orleans to learn about the communities, the disaster and the rebuilding. I have also brought together Ryerson and York University students to study the key issues — environmental degradation, history, race, class, disasters, and community development — as well as take part in assisting local communities in the New Orleans area for two weeks. While there we meet with community groups, activists and politicians, and tour the city to see the disaster first-hand.

Climate change debates are a definitive part of Katrina's legacy. Shortly after the storm a *New York Times* article stated:

> The hurricane that struck Louisiana and Mississippi on Monday was nicknamed Katrina by the National Weather Service. Its real name is global warming... As the atmosphere warms, it generates longer droughts, more intense downpours, more frequent heat waves, and more severe storms. Although Katrina began as a relatively small hurricane that glanced off southern Florida, it was supercharged with extraordinary intensity by the high sea surface temperatures in the Gulf of Mexico. (Gelbspan, 2005)

However, others point out that if climate change is the cause of increased hurricanes then there should be a larger number not just in the Gulf of Mexico but in all of the oceans; so far that hasn't come to pass (Lowry, 2005). For each person who blames climate change for Katrina's impact, there is another who shouts the opposite. Perhaps the debate can best be summed up as follows: "The chaotic nature of weather makes it impossible to prove that any single event such as Hurricane Katrina is due to global warming. It is also impossible to prove that global warming did not play a part, so debates about the causes of individual events are futile" (Young, 2007).

In this paper, I will argue that the climate change debate is of little relevance to the people most affected by Hurricane Katrina. Two other key issues — environmental degradation and environmental justice — are more clearly connected. If it had not been for the historically-developed vulnerabilities of the communities it affected, Katrina would not have been the disaster it was. Indeed, communities around the Gulf of Mexico aren't strangers to hurricanes. Hurricane season begins in June and doesn't wrap up until the end of November. Hurricane warnings occur regularly; most are small while others cause enough concern that people leave town for a few days. In New Orleans, until 2005 anyways, hurricanes were often an excuse for a party. The younger generation threw "hurricane parties" — giant neighbourhood potluck BBQs to use up whatever food may spoil if the hurricane knocks out the power for a few days. It was a celebration of life and survival, New Orleans style. The older generation remembers Hurricane Betsy from 1965; many of them stayed in town for it, lived through it and dealt with the aftermath caused by flooding and breached levees. Known as "Billion-dollar Betsy", she was the first U.S. hurricane to cause over $1 billion in damages; the flooding pattern was similar to Katrina but on a much smaller scale.

Hurricane Katrina made landfall at 6 am, August 29, 2005, in Buras, a small community about 100 km to the south-east of New Orleans. However, the strongest winds of a hurricane, because it spins counter-clockwise, are those on the west side. High winds from the storm pushed water up the Mississippi River Gulf Outlet channel, known as "The Mr. Go", which is a waterway designed to ease access to the city from the Gulf of Mexico. This water rushed into the local canals while water from previous built-over wetlands flowed into Lakes Borgne and Pontchartrain, and then pushed on into the city. Within hours, levees throughout the city had been breached and overtopped and flooding was everywhere.

The breach of the levees protecting the Greater New Orleans Area — especially near the parishes of Orleans and St. Bernard — caused massive destruction and devastation. Eighty percent of New Orleans had flooding, whether a few inches or several feet. All of St. Bernard Parish (a mostly rural community to the east comprised of 465 square miles in land area and over 1,300 square miles of wetlands) was under water; waters reached over 20 feet high in some places. Indeed, every house but two in St. Bernard Parish was declared uninhabitable (Spillman, 2009). The latter became the only community in North America to ever be 100% affected by a natural disaster. Hurricane Katrina was the costliest environmental disaster in U.S. history — estimates range around $200 billion — and with close to 2,000 directly-related deaths, there was definitely a clear financial and human toll.

In the immediate wake of Katrina, images on TV showed families, children, and the elderly at evacuation centres, left outside in the blistering sun with no water or no food. People were stuck on their roof tops, in their attics, hanging out a window, wading or being carried through toxic sewage-filled flood waters with bodies of dead people and animals floating by.

Notwithstanding these images, there were also more subtle and yet equally important consequences of Hurricane Katrina. The hurricane exposed and made clear what many call the "underbelly" of an American city; the ways in which race and class combine to produce vulnerable communities. Katrina also showed what happens when environmental degradation occurs in a systematic way; the flooding was a direct result of the destruction of the wetlands which serve as nature's barriers to flooding. The damage was also magnified by the environmental destruction that had occurred through the expansion of the city boundaries and the oil and gas extraction industries.

The city of New Orleans and the nearby parishes, in addition to being built on swamp land and in environmentally sensitive areas, are surrounded by levee systems. The levees were designed by the Army Corps of Engineers to reduce the yearly flooding that occurred when the Mississippi River overtopped its banks and later to protect the city from additional flooding. However, the levees have prevented the natural build-up of sedimentation that allows the wetlands to flourish and expand. Approximately one football field of wetlands disappears every 30 minutes; around 34 square miles a year. The city is also protected by a series of pumping stations that force the water out of the low-lying areas of the city; even a heavy rainstorm can cause a few feet of flooding in some of the lowest areas of the city. Nature's storm barriers have been destroyed because of the levee construction and especially by the oil and gas extraction industry which destroyed wetlands to build canals and pipelines (Figure 1). A traditional measurement, somewhat contested now, states that every three miles of wetlands reduces the storm surge by one foot (Masters, n.d.).

Figure 1: In Bayou Sauvage "elbows" from Cypress trees are all that is left after the storm tore through the largest urban wetland in North America. Once filled with trees tall enough to create a full tree canopy there is little now of the once majestic forest (photo by author).

New Orleans is a city that was built up because of its ideal location as a port economy. The nearby Gulf of Mexico and the Mississippi River provided easy access to much of the country. It was an agricultural stronghold in the American South with hundreds of plantations — using thousands of slaves — lining the river. Surrounded by water on three sides, as the city expanded it looked towards the wetlands; lack of government foresight allowed building to take place in extremely environmentally sensitive areas, and more levees were constructed to protect the newest areas of the city. St. Bernard Parish, bordering on the city's Lower Ninth ward sits on hundreds of acres of swamp land. It was formed primarily through "white flight" in the 1960s, as white New Orleanians moved out of the increasingly dangerous, poor and "black" city into the suburbs.

Katrina revealed a side of American life that is often hidden; race and class continue to be factors that limit people's ability to thrive and survive at the best of times, let alone during crises. The faces we saw on the TV were mostly black; when they weren't, they were the faces of lower-income white residents. Those with wealth and connections escaped the city before the storm hit. Elderly people or those with disabilities were often unable to make the journey by car and stayed (or were left) behind. These are also the people most affected by the storm and struggling to rebuild now — even five years later many are not back home.

Many people impacted by Hurricane Katrina continue to suffer from anxiety, depression, and stress. "Katrina has assaulted all the senses, and it is not over yet. This was not an acute injury, it is long term. It is not a posttraumatic stress disorder because we are still living it daily. One has the feeling that New Orleans is on life support and is struggling to survive" (St. Pierre, 2007). While there have been improvements in the three years since St. Pierre wrote these words, they continue to resonate today. The mere experience of living amongst abandoned houses, or seeing the remaining signs of the disaster — destroyed properties, lack of trees, search marks on buildings — creates a constant reminder of the tragedy and loss (Figure 2).

The images the world saw on the TV in the days following Katrina thus reflect the nickname given the city by residents — The City that Care Forgot. A few years prior to Katrina, Didier Cherpitel, Secretary-General of the International Federation of

Figure 2: An abandoned and destroyed house sits in the shadow of the Clairborne Street bridge in the Lower Ninth ward. This picture was taken in October 2009, over four years after the house was destroyed by Katrina (photo by author).

Red Cross and Red Crescent Societies, wrote, "In many cases, nature's contribution to 'natural' disasters is simply to expose the effects of deeper, structural causes — from global warming and unplanned urbanization to trade liberalization and political marginalization" (Jackson, 2006). That is certainly what happened in New Orleans.

Prior to Katrina, the 2004 U.S. Census reported that 67.3% of residents were African-American; the sixth largest black city population nationally. As a metropolitan area, it has the third largest black population in the U.S. at 37.5%. According to Sherman and Shapiro, "About one in every three people who lived in the areas hit hardest by the hurricane were African-Americans. By contrast, one of every eight people in the nation is African American" (Sherman and Shapiro, 2005).

New Orleans, despite the significant African-American population, experiences a spatialization of race and a racialization of space, which is the result of "decades of housing discrimination, environmental racism, urban renewal and police harassment"

(Lipsitz, 2006). Some areas of New Orleans approach homogeneity and are primarily African-American, such as the Lower Ninth Ward, which pre-storm was 98% black. This racial geography mirrors the highly damaged flood areas. The Lower and Upper Ninth Wards and Eastern New Orleans, all primarily African-American communities, were flooded significantly. Uptown, Bywater, Fabourg-Marigny and the French Quarter, all primarily Caucasian neighbourhoods, were minimally affected. This racial divide is mainly localized to the city of New Orleans. Flood waters followed no racial boundaries in St. Bernard Parish, a suburb created through white flight in the sixties. It was 93% white before the storm and 100% affected; flood waters knew no racial boundaries in St. Bernard. However, the racism that created the white flight mentality perseveres. The parish council attempted to pass a blood relative law which stated that properties could only be rented to blood relatives; thus intending that the parish would remain primarily white. This law was overturned in the courts but speaks to the continued racialization of space that continues today.

Redevelopment of New Orleans has been slow in some of the hardest hit areas. The "right to return" battle faces many challenges, including the financial cost of rebuilding, cheating contractors, slow government aid with complicated procedures, red tape, fraud, and shoddy workmanship and materials, and rezoning.

In St. Bernard recovery is slow all over and neighbourhoods throughout the parish face similar challenges (to New Orleans) in rebuilding, as well as the additional barrier of the lack of an economic base for the municipal government. Unlike New Orleans — which quickly regained its Central Business District, tourist areas and residential areas for the middle and upper classes, thereby resulting in a useable tax base — St. Bernard's recovering industrial areas cannot balance the significant loss of the commercial and residential tax base. Given the large number of damaged homes and businesses, recovery in the area is very slow — people need a house before they can move back home. But until people have a job, shopping areas, and neighbours, they won't build a house.

To make matters worse, St. Bernard Parish is now affected by the oil spill following the explosion on BP's Deepwater Horizon oil rig on April 20, 2010. As hurricane season gets underway, and the

Figure 3: Oil appears on both sides of the hard red boom near Grand Isle, LA in June 2010. The white soft boom is absorbent and soaks up the oil, while the red should keep oil out (but has failed to do so). This boom surrounds a barrier island that holds a brown pelican nesting ground (photo by the author).

oil continues to spread, there are fears about how bad it is going to get. Unlike a hurricane, from which a family or community can eventually rebuild, oil has the potential to completely change the way of life for these communities. Fishing and shrimping are more than a job, they are a way of life; children often work alongside their parents before obtaining their own boat.

Yet New Orleans is rebuilding, slowly, through the work of a variety of organizations, including Common Ground, Beacon of Hope, LowerNine.org, and the St. Bernard Project. The latter is a community-based rebuilding project that has brought over 275 families home. Working on 50 houses at a time, it costs $12-15,000 and takes ten to twelve weeks to rebuild a home. Supported by a diverse range of funders including Patron Tequila, United Way and individual donors, the St. Bernard Project also runs a Health and Wellness Center to help people cope with emotional health. These days they have expanded their operations to provide support to fishing families impacted by the oil spill.

As with Katrina, the challenge of responding to the oil spill has been taken up by non-profit organizations. I have just spent a week working with Catholic Charities at two relief centres (funded by BP) helping fishing families access food and supplies. I've also visited two of the Louisiana communities — Grand Isle and Venice — most affected by the spill. Both the Vietnamese-American and the Cajun communities there have been significantly affected by the oil spill and are afraid of what may happen and how their way of life may need to change. Many people that I met explained that they don't know how they will be able to recuperate from this latest tragedy. One shrimper told me that this was going to be the year the shrimp were plentiful, the price was high, and he was going to be able to get ahead. Instead, he waited in line, head bowed, eyes teary, to get a $100 grocery store voucher to feed his family.

The real name of the disaster that struck New Orleans is neither Katrina, nor climate change. It is instead vulnerability on the part of abused natures and exploited humans. Katrina and climate change hit those people who were the most vulnerable; typically racialized and poor communities. Rebuilding in New Orleans means starting over for many, and at the same time, cre-

Figure 4: A volunteer with the St. Bernard Project works to rebuild a home in Arabi, Louisiana for low-income seniors in October, 2009 (photo by the author).

ating some semblance of normalcy. Yet really, nothing will ever be the same. Families have been torn up, homes destroyed, communities flung across the country. And yet, there is a spirit, a perseverance that exists in New Orleans. It is like the popular food dish that combines whatever you have in the house, what local celebrity Phyllis Montana-Leblanc calls New Orleans "gumbo", no matter what goes in, something good will come out. So will it be with New Orleans, but only if there is a focus on socially and environmentally vulnerable communities and their particular needs and aspirations.

* * *

Tanya Gulliver *is a PhD student in Environmental Studies at York University exploring issues of vulnerability, disasters and resiliency with a particular focus on Hurricane Katrina. She has a background in social justice issues, and has worked as a university instructor, consultant for non-profit organizations and freelance writer.*

REFERENCES

Gelbspan, Ross, 2005. "Hurricane Katrina's real name," *New York Times*, 31 August.

Jackson, Stephen, 2006. "Un/natural disasters, Here and there." http://understandingkatrina.ssrc.org/Jackson/, accessed 15 September 2009.

Lipsitz, George, 2006. "Learning from New Orleans: The Social Warrant of Hostile Privatism and Competitive Consumer Citizenship," *Cultural Anthropology*, 21, 3, 451-468.

Lowry, Rich, 2005. "Katrina Conceit: Global warming and Mother Nature," *National Review Online*. http://old.nationalreview.com/lowry/lowry200508300805.asp, accessed 25 May 2010.

Masters, Jeffrey, n.d. "Storm Surge Reduction by Wetlands," *Weather Underground*. http://www.wunderground.com/hurricane/surge_wetlands.asp, accessed 20 May 2010.

Sherman, Arloc and Isaac Shapiro, 2005. "Essential facts about the victims of Hurricane Katrina." http://www.cbpp.org/cms/?fa=view&id=658, accessed 15 September 2009.

Spillman, A. 2009. Personal Communication.

St. Pierre, Jerry, 2007. "Be prepared: Katrina's aftermath — health care in crisis," *American Journal of Obstetrics & Gynecology*, 196, 6 (June), 561-563.

Young, Emma, 2007. "Climate Myths: Hurricane Katrina was caused by Global Warming," *New Scientist*. http://www.newscientist.com/article/dn11661-climate-myths-hurricane-katrina-was-caused-by-global-warming.html, accessed 16 May 2010.

JAY PITTER

Unearthing Silence
Subjugated narratives for environmental engagement

Narratives are an anchor that ground fact and theory in everyday life. They have the power to defy margins, educate, foster meaningful relationships and engage communities. As such, I am inspired by the work of individuals like Cruikshank (2000) and Poletta (2006), whose research explores the strategic use of narratives to advance social movements, honour diverse ways of "knowing" and safeguard human rights. As a Black, female environmentalist I am particularly concerned that the movement's dominant narratives fail to reflect the breadth of concerns and inter-connectivity of environmental and social issues.

Drawing on theory and lived experiences of narrativists and environmental thought-leaders, this paper will explore and interrogate the lack of subjugated environmental narratives conveying under-told and unauthorized experiences and knowledge. Using a narrative tracing the journey of a young woman and her mother back to Africa as a touchstone, the paper makes the case for unearthing subjugated narratives that are understandable, grounded in everyday life and mindful of social issues. It will argue that citizen environmentalists like the young woman in the aforementioned narrative, intuitively understand the inextricable connection between all living beings and their natural and social systems. The paper concludes by highlighting how the lack of diverse narratives not only creates a barrier to meaningful community engagement, but also impedes our ability to find solutions to an increasingly complex and broad range of environmental issues.

Environmentalism and its issues are at once extraordinarily broad and narrow. Formally conceived in the 1960s, the movement has defined a hierarchy of concerns such as climate change, pollution, sustainable forestry and protection of endangered species. These concerns or issues are conveyed through dominant environmental narratives, which embody the social norms, values and knowing of decision-making groups. Like dominant narratives across disciplines, they have emerged amid assumptions and exclusions within institutional and grassroots contexts.

According to Schoenfeld (2004), "the environmentalists' narrative is the story that the environmentalist community tells that answers fundamental questions about itself." Unfortunately, the vast majority of dominant environmental narratives tend to be:

- Discipline-specific and/or rooted in positivist science.
- Void of social context.
- Laden with scientific jargon.
- Communicated by highly credentialed experts.

Meppen and Bourke (1999: 389) accede and assert that, "conventional conceptualization of environmental problems has remained a largely disciplinary-based exercise that has relied on abstracting the environmental issues from their real-world complexity." Conversely, subjugated narratives like the following may challenge dominate narratives by creating space for diverse perspectives and knowledge:

Noelle struggles through tears while recalling a recent return to Rwanda and Ethiopia. She travelled with her mother whose life had been torn apart by rape, torture, and displacement resulting from multiple massacres and genocide of the Tutsi ethnic group starting in the late 1950s culminating in 1994. Together, the pair set out to confront history and locate Noelle's Ethiopian father.

Upon her arrival to Ethiopia, Noelle is struck by the way poverty is enacted in public spaces. The streets are carpeted with litter; entire families beg for money, men openly relieve themselves, and people bucket-bathe in plain view. This is a stark contrast to the time she had spent in Rwanda. While visiting the land of ten thousands hills, Noelle notes how post genocide policing and reconstruction resulted in pristine streets where plastic bags were illegal. It was almost as though the new government was attempting to sweep away the nation's painful past.

However, in Ethiopia the stench of social disparities rise from side street gutters accosting tourists patronizing jewellery stores and European-style bakeries. This is the irony of Ethiopia's capital, Addis Ababa, meaning beautiful new flower. This city of sharp contrast and contradictions is not only the site of Noelle's returning, but is also the context for a compelling environmental narrative about land-locked peoples, language, water-scarcity, memory and blood.

"My water broke while I sat in the back of a hired car. The driver was angered and insisted that I get out immediately. I knew that your birth would make my life a big mess." This is what Noelle's mother tells her when she is six years old. So three decades later when her mother sets out to locate the man she describes as "a forgettable sexual partner with other commitments," Noelle purchases a wooden Orthodox Ethiopian cross wanting to believe in something more hopeful than the stories she had been fed as a child.

Days later Noelle is invited to the meet her father's family. They pour over family photos and spin tales of a great and generous man who dearly loved all of his children, including Noelle. Amid the excited chatter, Noelle can't help but notice that all of her female cousins have numerous children. She wonders if like nutritious foods, warm water and education, these women cannot "afford" sexual autonomy. Quickly shifting her gaze Noelle begins to examine the privilege of choice, education and resources she had been granted growing up in Canada. Becoming increasingly anxious, Noelle blurts out the question that isn't being answered. "Where is he...my father where is he?" Receiving her into his arms, Noelle's uncle explains that her father had passed on over a decade ago but that he would be honoured to stand in his place.

In a dry nation, interpreted through a foreign tongue, Noelle feels known for the very first time.

What brought about the relaying of this experience was a simple Facebook message (and subsequent interview) issued on a Saturday morning, inviting folks to share an important environmental narrative with me. Like many narratives culled from historically excluded peoples, it exemplifies the inter-connectivity between environmental and social issues. It also anchors environmental issues in what David Herman (2009) might refer to as a "storyworld in flux" and in doing so creates meaning. Noelle's narrative challenges dominant narratives by both unifying and complicating topics such as forced migration, pollution and water-scarcity. She also calls attention to issues like rape, poverty and gender equity, which are not generally understood as being environmental. Though subjugated, Noelle's narrative is both compelling and credible because it is anchored in an actual experience, rather than filtered through a narrow discipline lens.

While her narrative does not provide statistical data or pose solutions for each environmental issue raised, it is an important addition to the dominant narrative because it has emotional resonance and contextualizes a number of important issues. "The authenticity of personal storytelling makes the form trustworthy — sometimes more trustworthy than the complex fact and figures offered by certified experts" (Polletta, 2006: 27).

In addition to being trustworthy, many subjugated narratives like Noelle's uncovers new ways of knowing that mediate distances such as generation, gender, language, class and historical hurts. Being alive in unrestrictive, open narratives may create meaning and connects complex issues like the ones environmentalists are contending with.

For example, when Noelle is waiting for her mother she recalls looking down at the war-stained earth, which begins to retell her mother's narrative of genocide and rape. Noelle imagines her mother fleeing with that very earth beneath her feet. She sees her mother's naked body, vulnerable against the earth after being raped by a gang of rebels. Noelle also recalls her mother telling her that the Rwanda massacres were partially fuelled by the uneven distribution of resources and perceived lack of land. Those who spilt the blood of their neighbours immediately acquired plots of earth. Noelle's willingness to be in the narrative's moment expands her understanding of land and what it symbolizes within a very different social context than the one she grew up in. Noelle also recognizes the courage it took her mother to accompany her on the search for her father. Within that profound narrative moment, Noelle contemplates issues that are at once deeply personal and political.

When properly applied and received subjugated or alternative narratives facilitate critical and responsible political deliberation and participation (Stone-Mediatore, 2003). However, the absence of subjugated narratives like Noelle's prevents diverse communities from contributing to larger environmental discourse(s) and participating in the movement. Also, it is evidence of the continued exclusion and/or misrepresentation of historically marginalized groups.

Fortunately, issues of knowledge, power and voice are now entering environmental sciences and challenging master scientific narratives (Cruikshank, 1998). An emerging number of environmentalists like Gosine and Teelucksingh (2008) are advocating for and using an environmental justice framework for addressing the complexity of the issues. This framework enables environmentalist to not only identify and monitor environmental hazards, but also encourages strategies that transform ideologies of race, gender and class to ensure safe physical and social structures for everyone (Hogan, 2004).

Through the diverse themes raised in Noelle's narrative, it is clear that she too is conscious of the inter-connectivity of environmental hazards and social location(s). When re-telling her experience in Ethiopia she begins by highlighting the huge class gap between the locals and visitors like herself. She notes the contrast and contradiction between fine European style bakeries and entire families begging for food. She also voices concern for her female cousins. Without judgement or recrimination, she wonders how poverty resulting from water scarcity, ongoing war and an unstable government has impacted their ability to exercise sexual and reproductive agency. Noelle's recognition of systemic power, observation of class and gender along with her willingness to interrogate her values suggests that she is intuitively processing from an environmental justice framework.

Hearing and analyzing Noelle's story using an environmental or social justice framework is imperative because core causes of environmental threats are not strictly due to our abuse or negation of the natural world and non-human beings; but rather in our on-going abuse and negation of each other. The questions aren't whether or not we have enough clean water to sustain human life, or how quickly the earth's temperature is rising or if we love polar bears. They are much more serious and complex.

They centre on whether or not we are prepared to share the earth's resources, respect diverse ways of knowing, discard old paradigms of power, forgive historical wounds and create spaces where all narratives are heard and a diverse range are authorized.

In presenting Noelle's narrative I am not asserting that subjugated environmental narratives are homogenous or that they should be privileged over scientific dominant narratives; but rather that they should inform them. I am strongly suggesting that the inclusion of both environmental and social issues in subjugated narratives would expand and complicate the environmental discourse. This would likely contribute to better engaging communities and devising holistic and sustainable strategies for addressing environmental issues.

* * *

Jay Pitter is a student in the Master of Environmental Studies Program at York University. She is also an accomplished marketing communications specialist and writer with numerous credits including the Walrus, Toronto Star, Fireweed, *Sister Vision Press, World Stage, TVO, Vision Television and CBC Radio.*

REFERENCES

Cruikshank, J., 2000. *The Social Life of Stories: Narratives and Knowledge in the Yukon Territory*. Lincoln: University of Nebraska Press.

Gosine, A.and Teelucksingh, C., 2008 *Environmental Justice and Racism in Canada*. Toronto: Edmond Montgomery.

Herman, D., 2009. *Basic Elements of Narrative*. United Kingdom: Blackwell.

Hogan, K., 2003. "Detecting Toxic Environments: Gay Mystery as Environmental Justice," pp. 249-61. In Rachel Stein, editor, *New perspectives on environmental justice: gender, sexuality, and activism*. Piscataway, NJ: Rutgers University Press.

Meppen, T. and Bourke, S., 1999. "Different Ways of Knowing: a communicative turn toward sustainability," *Ecological Economics*, 30, 3, 389-404.

Polletta, F., 2006. *It Was Like A Fever*. Chicago: University of Chicago Press.

Schoenfeld, S., 2002. The Environmentalists' Narrative (paper). Environmental Studies Association of Canada. http://www.sustreport.org/resource/shoenfeld.html, accessed 15 June 2010.

Stone-Mediatore, S., 2003. *Reading Across Borders: Storytelling and Postcolonial Struggles*. New York: Palgrave Macmillan.

Part III

Beyond Climate Change and Chilly Climates

ELIZABETH MAY

A Practical Environmental Education
Shrinking ecological footprints, expanding political ones

When I was in elementary school, decades ago, there was no concept of environmental education. We learned Biology and Chemistry and often the examples used to explain the subject came from nature. By the time I was in high school, in the 1970s, I started an ecology club in my school. Sympathetic teachers backed us up. We leafleted on the first Earth Day, April 22, 1970. We called for banning phosphates in detergents, with my chemistry teacher overseeing our experiments, so we could measure the pH of various commercial laundry soaps. Environmental education was in its infancy, but already important.

In recent years, the topic has been "in" and then "out" of various provincial curricula. Over the last few decades, I have often been invited to speak to school groups — elementary to high school — as well as to environmental educators. Much progress has been made with tremendous leadership from dedicated teachers.

Within the range of human activities with the adjective "environmental," the aims and benefits of environmental education are generally presented as safe, non-threatening action. Who could object to educating our young about the perils facing our planet?

Yet, as I prepared to write this article, what sprang to mind was a set of objections to certain approaches to environmental education.

What we don't want

THE "GENERATIONAL COP OUT" APPROACH

The first and primary objection is to what I'll call the "generational cop out" approach. It is often presented as the following: "The only solution to the environmental crisis is to change our cultural values, to achieve a paradigm shift. And we can only do this by educating our youth to embrace a different kind of relationship with the natural world. Our children will do a far better job than we have."

The problem with this statement should be obvious: it lets the generations with decision-making power off the hook and assumes we can leave the current environmental crisis to our children. That is something which we most emphatically must not do. The climate crisis requires meaningful, aggressive, desperately ambitious action to slash greenhouse gas emissions within the next seven years. Please read that last sentence again. According to the *World Energy Outlook 2007* (International Energy Agency, 2007), global emissions of greenhouse gases must peak, stop rising, and begin a descent no later than 2016 in order to avoid runaway global warming, in which the impacts would trigger an unstoppable and destabilization of the climate system through positive feedback loops.

Environmental education should not let us (anyone over 18 and with breath of life) off the hook. We must all demand dramatic changes in our policies and our actions.

The upcoming United Nations negotiations in Copenhagen this December are our last chance to successfully negotiate the requisite emission reductions. The Copenhagen goal is to develop a new treaty meant to kick in when the first phase of Kyoto ends in 2012. This time all industrialized nations have to commit to meaningful action and deliver on it. Developing countries will have to commit as well, although with a slightly different approach.

Under the Harper government's policies, Canada is sabotaging progress. Fortunately, we are no longer respected in the world on the environmental front so few countries listen to us. Our national delegation is in the way, arguing for global "intensity based targets," defending the tar sands at every opportunity, and trying to forge alliances with any back-sliding nation it can find. It is heartbreaking that a country with a once envied reputation for global leadership in peace-keeping, and respect for human rights and environmental multilateralism, is now ignored. It is even more heart-breaking to admit that it would be far worse for the world if we had much influence.

CORPORATE-BRANDED APPROACH

The second approach to environmental education to which I raise objection is the corporate-branded kind. Learning about forestry in New Brunswick from the Irvings or about nuclear power in Ontario from Atomic Energy of Canada Ltd. are two real examples of industry self-promotion through our schools.[1] Now that schools are so desperate for cash that they can be bought off by Coca-Cola, no one should be surprised when industry propaganda sneaks in to augment school materials. This approach, like the Coke and Pepsi contractual monopolies, must be rejected (Manning, 1999).

THE FEAR APPROACH

The third objection is equally passionate: we must not provide an environmental education that paralyzes our children with fear. This challenge is the toughest. As my first objection suggests, we are nearly out of time to avoid a list of climatic events for which the word apocalyptic was invented. The most recent science from one of the top research facilities in the world, the Hadley Centre in the UK, estimates that if even all countries reduce greenhouse to targets that only the most ambitious nations have proposed, the world will still only have a 50-50 chance of avoiding runaway global warming.

Now, as a clear-eyed realist, I will take that 50-50 chance. I will embrace it for what it is — a reasonably good chance at survival for human civilization. We have delayed, procrastinated, denied and debated and blown any chance at a sure thing. We no longer have the prospect of a guaranteed success once (or if) serious action is taken.

Back in 1986, when I first started working on this issue, we had that chance. We had it at the Earth Summit in Rio in 1992. But by the year 2000, with greenhouse gas levels continually growing globally (even as the European Union met its Kyoto reduction targets), the U.S. and Chinese pollution rates, with Canada polluting well above its per capita "fair share," have deprived us of any hope of avoiding serious climatic disequilibrium.

The challenge for responsible educators is how to explain the science, its chemistry, and its politics to children and young adults in a way that does not terrify them. Environmental education is critical. The message about the threat posed by climate change or toxic chemical contamination must be conveyed with a large focus on solutions. No young person should be taught about current environmental challenges without first acquiring a solid historical base of successes in the environmental movement. A short and incomplete list would include learning about how citizen pressure resulted in the end of atmospheric nuclear testing (arguably the first global toxic contaminant) through the Test Ban Treaty in 1963, how Rachel Carson wrote *Silent Spring* and DDT was banned, how lakes and rivers returned to health once phosphates in detergents were banned, how the Montreal Protocol worked to protect the ozone layer, and how agreements to cut emissions of sulphur dioxide between Canadian provinces and the federal government worked to push the U.S. to agree to do the same, largely eliminating the damage done by acid rain.

We need to convey to young people that our current problems are not insoluble. We need to stress the solutions to every problem examined.

These young people, even the very young, know a lot of what is going on. They do not want to be kept in ignorance, but the truth of our situation must be conveyed in a way that empowers them to be full and productive citizens. One Grade 5 student in Nova Scotia asked me after a class presentation, "What happens after global warming?" I tried for a simple explanation — honest, but not horrifying. I said, "It depends. It's like asking what happens if a car hits a brick wall? If your dad's foot is on the brake and he is slowing down a lot, then it might not even dent the bumper, but if the car is not slowing down..." A little boy interrupted to ask in an alarmed tone, "then, we all die?"

Quickly I re-assured them. "The whole world is putting the foot on the brakes," I told them, "we can still stop before we hit that wall." As I said it, I knew that humanity's heavy foot is still on the gas. To grown-ups, I repeat, we cannot leave this to our children.

Just as much as we cannot ignore the crisis until they come along to take over from us, neither can they arrive in adulthood without the tools of effective citizenship.

What we need in environmental education

There are four quite distinct but important requirements of a practical environmental education.

GREENING THE CURRICULUM

The first starts with "education" at its most conventional, bringing relevance to the standard curriculum. The educational opportunities presented by environmental issues apply across many core areas of the curriculum from elementary through to high school (and for that matter, to life-long learning). There are elements of biology to be understood in ecosystems, food chains, ecological niches, photosynthesis, and so on; of chemistry in terms of toxic chemicals and their differing impact due to locations of receptors for bonding, bio-chemistry, the chemistry of the atmosphere, acid rain and sulphur dioxide creating sulphuric acid, carbon dioxide and other greenhouse gases, and so on; in physics for understanding production of electricity from different sources, for understanding how nuclear power works, and so on; for world issues: the issues of equity, of developing-country challenges when faced with famine, persistent poverty, and the impact of market forces, dumping, tariff barriers, the WTO all the way to multilateral agreements and the UN system, the Montreal Protocol to protect the ozone layer, and so on. The list of useful immediate applications of environmental education to a wide array of course work is nearly unlimited to a creative educator.

EDUCATING A RE-ACQUAINTANCE WITH NATURE

Of equal importance is the environmental education that acquaints the child with the natural world. The ability to name at least as many species of plants and birds as of clothing and fast food brand icons is a survival skill. Knowing nature as a friend, pondering nature as a vast mystery, allowing nature to stir the soul and inspire the creative potential of the future artist and poet, is under-rated in our education system.

Renowned Canadian artist Robert Bateman has written about the urgent need to address the "nature deficit disorder" in our children:

> From the beginning of time we have been connected to nature. We, of course, are literally nature's children, but for the first time in history that connection threatens to be broken by the majority of an entire generation and perhaps generations to come. When I read Richard Louvs' landmark book, *Last Child in the Woods — saving our children from nature deficit disorder*, I was surprised at how insidious and widespread the problem had become…. Most children are not playing by themselves out in nature. Almost all outside activity is adult supervised. Soccer moms are a relatively recent phenomenon but adult supervision seems to be essential nowadays…. *Last Child in the Woods* cites recent research at Harvard and numerous U.S., Canadian, British and European institutions. The findings are that if children play in nature, (I don't mean organized soccer or cement playgrounds), climb trees, build forts and dams in creeks and go exploring, here is what happens: they have less obesity, less attention deficit disorder, less depression, less suicide, less alcohol and drug abuse and less bullying and higher marks. If one was to make a list of the main problems facing that age group and indeed any age group, it would be the same list. And nature is free. (Bateman, 2009)

There are a number of ways to effectively integrate nature into a school's agenda. Invite a local naturalist to take school groups around any block while explaining to the children how to identify a tree by its leaf, the shape of its tree branches and its bark, letting them sketch and trace leaves. Have a competition within the class to identify the largest number of plants and flowers. Anywhere in Canada, even in an urban area, a walk around the block with a knowledgeable naturalist will open a new world to students.

Another area of critical importance, as Robert Bateman suggests, is unfettered, unregulated play time in nature. While there is little schools can do about a child's weekends and after school, it is possible to naturalize playgrounds, to plant school gardens, to encourage a greener experience. Here, again, the evidence is clear. Children playing on grass, rather than asphalt, children able to swing from a tree branch rather than a wire mesh fence, are less aggressive and become better learners. (The collateral benefits of more carbon sequestration, less run off from rainfall to storm sewers, and a real understanding of where food comes from are worth mentioning as well.)

In the area of practical knowledge are the life skills for a smaller ecological footprint. One of the more unheralded ways to live more lightly on the land is to know how to prepare a meal from basic raw (healthy) ingredients. Increasingly, kids don't watch a parent cook a meal at home, and if they do, a good portion of that meal may be in the form of "packaged, convenience" foods. Busy and hectic schedules and the advent of single service microwaveable meals mean that a lot of kids know how to heat up junk, but may have no idea how to make a healthy meal. As schools have (not entirely, but frequently) ended home economic courses, young people head off to university or college without a clue how to cook. Post-secondary institutions are increasingly in the clutches of corporate monopolies providing meal plans in cafeterias. Worse, many have turned student common areas into replicas of "food courts" in malls, lined with fast food creating mountains of throw-away containers and future ill-health.

Schools, as part of an environmental education, need to start some life skill classes. What used to be called "shop" could be "Reduce throwing things out: How to repair your stuff," "How to make your own furniture." With a naturalized school ground, "How to grow your own garden" could become an element of course work. These topics are gender neutral, just as "How to eat for a healthy planet" could attract both boys and girls. The fringe benefit of knowing how to make excellent meals and save money by buying real ingredients and whipping up something healthy and delicious is just icing on the cake. (Yes, cakes from scratch help the planet too.)

THE IMPORTANCE OF ENVIRONMENTAL CITIZENSHIP

The last aspect of environmental education is learning the habits of effective citizenship. Once students know there are solutions for the environmental challenges we face, they need to know how change happens.

It is never too early to start writing letters concerning issues kids care about. The campaign to get endangered species legislation in Canada started in 1994 when Environment Minister Sheila Copps received thousands of letters from school children. It made a difference when I was growing up that a letter I wrote to a politician would be answered and that sometimes one of my letters would be published in the local newspaper. Depending on the age level the discussion of our political system, learning how Canadian Parliamentary democracy is supposed to work can be brought alive by including critiques on the system and on current politics. Too much of what passes for Civics in high school classes is merely rote learning of Canadian government as if it were an organizational chart. Young people are not voting, at least in some measure, because the relevance of the governmental system escapes them. How to force the government to change policy on how rivers get cleaned up and dangerous chemicals stop being used is just not clear to many young people.

Empowering students to see themselves as change agents is often reinforced through school projects. A school-based campaign, with a campaign plan, some media coverage, a pamphlet or poster to build up support in the student population can create much-needed confidence for later campaigns with bigger targets. Anti-idling campaigns, reducing energy use in the school, fighting to change a school policy (pesticide application, waste of paper through single-sided copying) can create an early success on which future successes in all aspects of their lives can be built. One frustration with high school environment clubs is the extent to which committed students spend their spare time sorting compost and recycling from garbage in order to maintain the programme. The best way to engage kids is to let them do more than volunteer janitorial work in the school.

Of course, any one change will bring to mind the need for larger changes. Much of what I have described here requires more funding for the schools, more time for teachers to help support student-run projects, and funding to rip out the prison-like asphalt and chain link fences to replace them with sod and trees and small hills to roll down. Society as a whole needs to be engaged to ensure quality education. An African proverb holds that it takes a village to raise a child. The whole needs of the child suggest to my mind that it takes a planet to educate them.

* * *

Elizabeth May, O.C. *is the Leader of the Green Party of Canada.*

ENDNOTE

[1] Irving forest programs are on their website: http://www.ifdn.com/.

REFERENCES

Bateman, Robert, 2009. Essay, "Children and Nature," March 09, unpublished, used with permission.

International Energy Agency, 2007. *World Energy Outlook 2007.* http://www.iea.org/textbase/nppdf/free/2007/weo_2007.pdf, accessed 1 July 2010.

Manning, Steven, 1999. "Students for Sale – How Corporations Are Buying Their Way into America's Classrooms," *Education Policy Studies Laboratory.* http://www.asu.edu/educ/epsl/CERU/Articles/CERU-9909-97-OWI.doc.

ALI LAKHANI, VANESSA OLIVER,
JESSICA YEE, RANDY JACKSON &
SARAH FLICKER

"Keep the fire burning brightly"
Aboriginal youth using hip hop to decolonize a chilly climate

Aboriginal youth & HIV in Canada

Members of Aboriginal communities are overrepresented in Canadian HIV/AIDS statistics, with an infection rate three times that of non-Aboriginal people (Public Health Agency of Canada, 2006). Canadian Aboriginal youth, women, and injection drug users are more likely to become infected with HIV/AIDS than non-Aboriginal populations (Ricci et al., 2008). The median age for HIV infections in Aboriginals in Canada has decreased from 32 years of age to 23 (Isaac-Mann, 2004).

HIV/AIDS infections tend to disproportionately affect the marginalized (Farmer et. al, 1996). One explanation for the elevated prevalence of HIV/AIDS within Aboriginal communities is that the ongoing systemic colonial oppression faced by Indigenous populations propagates condition of risk (Gracey & King, 2009; King, Smith & Gracey, 2009; Smyllie, 2009). Specific high-risk behaviors — including substance abuse, violence, and limited use of, and access to, healthcare services — are associated with, and exacerbated by, generations of violent colonization and residential schooling, and contribute to these elevated rates (Mehrabadi et al., 2008; Pearce et al., 2008). According to Gomes et al. (2004), local HIV epidemics are exacerbated by climate change which can impact relationships to land, agriculture, food security and the need for mobility. In the case of Aboriginal communities in Canada, climate change associated with industrialization has continued the colonial legacy disrupting Indigenous connections to the land and has fueled migration to urban centres.

Many of the conventional HIV prevention strategies that focus solely on individual behaviour and fail to take these legacies into account have proven ineffective (Flicker et al., 2008). Aboriginal youth struggling with histories of mistreatment and abuse require unique strategies to overcome these barriers, strategies that leverage their strengths and interests rather than their shortcomings. Studies show that HIV prevention programming that involves the reclamation of history and culture may work to challenge the racist stigmas attached to HIV and AIDS and other poor health outcomes (Hare & Villarreul, 2007; Morgan & Freeman, 2009, Evans-Campbell, 2008). This challenge can help Aboriginal persons reverse the self-blame currently prevalent in their lives (Larkin et al., 2007). Furthermore, peer programming that builds the capacity of youth to engage their peers in sexual health has proven to be successful (Maticka-Tyndale, 2006). The goal of *Taking Action: Using Arts-Based Approaches to Develop Aboriginal Youth Leadership in HIV Prevention* is to draw on youth talents and provide young people with a forum to become change agents and peer leaders in their communities (Flicker & Jackson, 2008). This paper describes how youth participating in our workshops used hip hop as a vehicle for talking about their environments, which they often describe as "cold" and "uninviting," in their efforts to curb HIV in their communities.

Origins of hip hop & Aboriginal practice

Hip hop is widely believed to have originated in the Bronx during the 1970s (Akom, 2009). This cultural practice emerged as a response to the oppressions faced by the African American and Caribbean communities living within postindustrial America (Rose, 1994). Hip hop was a "communal" activity: people would congregate and interweave aspects of art, performance, and dance into a form of expression. Today, hip hop has become a dominant language of youth culture; De Leon (2004: 1) argues that "those of us who work with young people need to speak their language."

Hip hop is often initiated by communities experiencing oppression. Indigenous groups around the world have led the way. Aboriginals in Australia and New Zealand have used the form to question Western domination and identity (Maxwell, 2002; Mitchell, 2001). Numerous Indigenous groups throughout Canada are also expressing themselves through hip hop.[1]

The histories of European colonization shared by Black and Aboriginal groups provides one explanation for this musical tradition to have "crossed-over" (Potts, 2006; Foreman, 2002). It is unsurprising, then, that Aboriginal youth are carving out identities through the resistance of hip hop culture. Further, there are strong ties between the music of Aboriginals and African Americans and shared roots of Indigeneity. This may be another reason why so many Aboriginal youth identify closely with the cultural practices of hip hop (Hollands, 2004).

The *Taking Action* workshops

Taking Action works with Aboriginal communities across Canada to build youth leadership, work on decolonization strategies, and create awareness around HIV prevention. We give youth an opportunity to pick the art forms they want to use to express themselves, and then link them up with local artists (where possible) who can help them amplify their message. In this paper, we reflect on two *Taking Action* workshops in which youth chose hip hop as a vehicle for talking about HIV in their communities. The Toronto group involved three urban Aboriginal youth. The Kettle and Stony Point workshop included 10 participants, all but one living on the same reserve. Both were weekend workshops that involved interactive HIV education activities, decolonization discussions, Elder support, and cultural empowerment activities, as well as facilitated songwriting and recording. A month later, each youth was interviewed individually about his/her experience.

Hip hop and decolonization

Workshop participants created songs about the inequities they face and discussed strategies to overcome them. The songs express the often harsh realities faced by the youth.[2] Youth predominantly viewed the numerous worlds that they navigate between as negative. For example, in "Do it Right,"[3] youth describe their surroundings as *troubled* and *unhappy*. "The Pain"[4] focuses on feelings of entrapment and escape mechanisms:

> The pain it feels so plain, I'll be sitting in my room and I'm going insane/ Smoke too much Jane, you know it man, I got nothing more to gain.

When asked to interpret what they were trying to say, one participant responded:

> Um, that they feel pain, so I guess to get rid of it they smoke weed. And they don't want to but I guess that's all they have so they just do it. (Interview 13) [*Many participants reiterate the point.*]

However, both songs also speak back to their harsh environments. The Toronto youth provided motivational statements to listeners encouraging them to *hold on* and *stand strongly* during times of trouble and misunderstanding. The participants assert that resilience and strength are necessary to combat the challenges in their daily lives. The bridge at the end is we need to stay united. In Kettle and Stony Point, youth demanded: *Hope and no more Dope!* The song, which protested the lack of extracurricular opportunity, brought to light issues of boredom and a dearth of social programming for the youth on the reserve. During post-interviews, youth explained:

> It is about all the young people [who] just do dope around here, so instead of that, like make a skateboarding park or something. Like, um, somewhere where you can play video games, like an arcade. We used to have one, but they closed it down. (Interview 15)

Many respondents shared this similar view and underscored the need for more recreational programs.

Participants were excited about disseminating their message to their peers and community. In their interviews, many reported that they had e-mailed the song to their peers and played it for friends using portable mp3 devices:

Yeah, I have the song on my computer. I like sent it around in an email to everybody that didn't get it. And then my buddy […] I sent it to him. He put it on his i-pod and showed it to everyone around the school, which is pretty cool. (Interview 16)

Participants also spoke about additional innovative techniques to get the music out to the greater community; for example, sending the song to the local radio station (Interview 15) and creating a website about the song (Interview 17).

Youth used hip hop music to affirm their identities and challenge those imposed upon them throughout the colonial process. Both groups' creations contain examples of hip hop being used as a mechanism to express decolonizing messages. The second verse of "Do it Right" is a good example:

They don't understand me, my jeans may be baggy/ that don't give the cops a right to grab me/ Unhand me.

In this instance, "they" refers to society generally and to the police specifically. The lyrics establish that the singer's identity is being misunderstood: his wearing baggy jeans draws scrutiny and harassment by police. He then challenges the system, police, or colonial representation, arguing that they have no right to grab him or treat him as a second-class citizen. He concludes the verse by lashing out at the oppressive forces, demanding *Unhand me*. The participant's critique of, and response to, police action, can be seen as a method of decolonization. The police represent one aspect of a colonial governing system that has sought to impose hierarchy and to destroy Indigenous culture. Therefore the participants' lashing out and arguing against the police through their lyrics is one way of "…thoroughly challenging the colonial situation" (Fanon, 1963: 3).

Similarly, the final verse of "The Pain" addresses the injustices faced by Indigenous peoples, and affirms that Aboriginal nations were the first to inhabit Canada:

sometimes we feel like we gotta run away/ but we know we gotta stand and fight the pain/ show everyone how we run this place/ First Nations people, it's like we got slapped/ straight across the face, see us you know/ we were here first.

The lyrics first describe the pain felt by the youth within First Nations communities and how people there must fight against that pain. The youth underscore that First Nations communities in Canada were maltreated, and provide the metaphor (if not a literal reference) of the First Nations communities being "slapped" by governing or colonial parties. The lyric concludes with the statement we were *here first*, powerfully affirming the original inhabitation of Canada by Indigenous people. Colonization has damped the Aboriginal voice, suppressed their culture and significantly reduced their land. In this lyric, youth are challenging the muting of the Aboriginal voice by speaking out in hip hop music. The process itself may be seen as decolonizing, as youth openly challenge the authorities that appropriated their lands, lives, and cultures.

Remarkably, HIV is not mentioned explicitly in either song. Both groups tried to capture in their music the underlying issues that lead to elevated risks of HIV without being overly direct. In follow-up interviews, participants who described a connection between their musical creation and HIV/AIDS noted how the actions and activities presented within their song, such as drug use and histories of colonization, contributed to higher rates of contraction. The Kettle and Stony Point drew a connection between drug use and HIV/AIDS:

> Something to do with like being high or something and then you might have sex. Yeah. And then they won't even know what they did, or something. [If they weren't high] they would have said no or something. Wore a condom. (Interview 5)

Reflecting on the meanings of their hip hop musical creations provided an opportunity for participants to identify factors that may contribute to HIV & AIDS, and their ability to identify these factors exemplifies how hip hop may be used as a tool in future HIV prevention programming.

Conclusion

Aboriginal youth participants in *Taking Action* workshops are using hip hop to express their realities, identify issues of concern, and discover solutions to those issues. They are challenging the colonial identities imposed upon them, while reclaiming

Aboriginal youth culture and identifying contributing factors to HIV/AIDS. Their use of hip hop represents the dynamism of Indigenous knowledge: while drawing on traditions, youth are creating "new" Indigenous knowledge centred on contemporary socio-political realities. We found the method to be particularly helpful in opening up a dialogue about the links between HIV and colonization. Other HIV educators and health promoters may want to think about how they can incorporate hip hop language, music and culture as a strategy for engaging youth in these challenging discussions. Our experience has been that when given the opportunity to express themselves, youth have stepped up.

That Aboriginal youth are using hip hop to reclaim their voice is encouraging in light of current and historical genocidal policies that have worked to eradicate Aboriginal voice and thought. Given the cultural erasure and incidences of trauma that are the result of residential schools and the colonial legacy, hip hop allows youth to reclaim their voice, give words to the injustices they face, and gain pride in their cultures and in themselves. It is not therefore surprising that so many Aboriginal youth have embraced this musical tradition rooted in resistance.

The growth of the urban music industry in Canada and Aboriginal forms of hip hop helps to create a climate that is more youth-friendly and ensure that Aboriginal voices finally "get repped".[5]

* * *

Ali Lakhani *is a recent graduate of the Masters in Environmental Studies programme at York University. He is a hip hop artist and film maker who is interested in the possibilities of using music for community development.*

Vanessa Oliver *is an Assistant Professor in Sociology at Mt. Allison University. Her research focuses on youth, sexual health and arts-based methods.*

Jessica Yee *is a proud Two Spirit young woman from the Mohawk Nation. She is the founder and Executive Director of the Native Youth Sexual Health Network — the only organization of its kind in North*

America by and for Native youth working within the full spectrum of sexual and reproductive health.

Randy Jackson *a doctoral candidate at McMaster University in the School of Social Work and a recent Ontario HIV Treatment Network Scholar. Originally from Chippewas of Kettle and Stony Point First Nation, Randy has been involved in a number of research projects that engage the community and incorporate Aboriginal values and perspectives.*

Sarah Flicker *is an Assistant Professor of Environmental Studies at York University and an Ontario HIV Treatment Network Scholar. Her research focuses on sexual health promotion with youth.*

Acknowledgements: *We would like to thank the rest of the Taking Action team: June Larkin, Claudia Mitchell, Tracey Prentice, Jean-Paul Restoule, as well as Sourav Deb and all the youth and elders involved. This research was made possible by grants from the Canadian Institutes of Health Research and the Ontario HIV Treatment Network.*

ENDNOTES

[1] Examples include groups such as Rez Official and 7th Generation; see also: http://www.beatnation.org

[2] See appendix for the lyrics of each song or visit www.TakingAction4Youth.org to hear the recordings.

[3] Song written during the Toronto workshop.

[4] Song written during the Kettle and Stony Point workshop.

[5] 'repped' – hip hop slang for the "represented".

REFERENCES

Akom, A., 2009. "Critical Hip Hop Pedagogy as a Form of Liberatory Praxis," *Equity and Excellence in Education*, 42, 1, 52-66.

De Leon, A., 2004. Hip Hop Curriculum: A Valuable Element for Today's Afterschool Programs. http://www.afterschoolresources.org/directory/promising-practices/programs.html, accessed 21 June 2010.

Evans-Campbell, T., 2008. "Historical Trauma in American Indian/Native Alaska Communities: A Multi-Level Framework for Exploring Impacts on

Individuals, Families and Communities," *Journal of Interpersonal Violence*, 23, 3, 316-338.

Fanon. F., 1963. *The Wretched of the Earth*. New York: Grove Press Inc..

Farmer, P., Connors, M & Simmons, J., editors, 1996. *Women, Poverty and AIDS: Sex Drugs and Structural Violence*. Monroe: Common Courage Press.

Flicker, S, Larkin, J, Smillie-Adjarkwa, C, Restoule, JP, Barlow, K, Dagnino, M., Ricci, C, Koleszar-Green, R., & Mitchell, C., 2008. "It's Hard to Change Something When You Don't Know Where to Start': Unpacking HIV Vulnerability with Aboriginal Youth in Canada," Special Issue of *PIMATISIWIN: A Journal of Indigenous and Aboriginal Community Health*, 5, 2, 175–200.

Flicker, S., Jackson, R., 2008. "Taking Action: Using Arts-Based Approaches to Develop Aboriginal Youth Leadership in HIV/Prevention." Submitted May, 2008 for the Operating Grant: HIV/AIDS (Aboriginal Community Based Research – Proposal 180571).

Foreman, M., 2002. *The "Hood' Comes First: Race, Space and Place in Rap and Hip Hop*. Middletown: Wesleyan University Press.

Gomes et al., 2004. Climate Change and HIV/AIDS Response: A hotspot analysis for early warning rapid response systems. United Nations Development Program. http://www.fao.org/forestry/15532-0-0.pdf, accessed 28 May 2010.

Gracey, M. & King, M., 2009. "Indigenous Health Part 1: Determinants and Disease Patterns," *The Lancet*, 374, 65-75.

Hare, M. & Villarruel, A., 2007. "Cultural Dynamics in HIV/AIDS Prevention Research Among Young People," *Journal of the Association in Nursing AIDS Care*, 18, 2, 1-4.

Hollands, R. 2004. "Rappin' on the Reservation: Canadian Mohawk Youth's Hybrid cultural Identities," *Sociological Research Online*, 9, 3.

Isaac-Mann, S., 2004. *Development of a Community-based HIV/AIDS Prevention Program for Urban Aboriginal Youth*. Unpublished Dissertation/Thesis.

King, M, Smith, A. & Gracey, M., 2009. "Indigenous Health Part 2: Indigenous Health Part 2: The Underlying Causes of the Health Gap," *The Lancet*, 374, 76-85.

Larkin, J., et al., 2007. "HIV risk, systemic inequities and Aboriginal youth: Widening the circle for prevention programming," *Canadian Journal of Public Health*, 98, 3, 179-82.

Maticka-Tyndale, E., 2006. "Evidence of youth peer education success." In *Youth Peer Education in Reproductive Health and HIV/AIDS*. Youth Issues Paper 7. Arlington, VA: Family Health International (FHI)/YouthNet.

Maxwell, I., 2001. "Sydney Stylee: Hip-hop Down Under Comin' Up." In *Global Noise: Rap and Hip-hop outside the USA*. Mitchell, T. (Ed.) Middletown, CT: Wesleyan University Press.

Mehrabadi, A., Craib, K., Patterson, K., Adam, W. et al., 2008. "The Cedar Project: A Comparison of HIV-Related Vulnerabilities Amongst Young Aboriginal Women Surviving Drug Use and Sex Work in Two Canadian Cities," *International Journal of Drug Policy*, 19, 159-168.

Mitchell, T., 2001. "Fightin' da Faida." In Mitchell, T., editor, *Global Noise: Rap and Hip-hop outside the USA*. Middletown, CT: Wesleyan University Press.

Morgan, R. & Freeman, L., 2009. "The Healing of Our People: Substance Abuse & Historical Trauma," *Substance Use & Misuse*, 44, 1, 84-98.

Pearce, M., Christian, W., Patterson, K., Norris, K., et al., 2008. "The Cedar Project: Historical Trauma, Sexual Abuse, and HIV Risk Among Young People who use Injection and Non-injection Drugs in Two Canadian Cities", *Social Science and Medicine*, 66, 2185-2194.

PHAC, 2006. *HIV/AIDS Epi Updates, Surveillance and Risk Assessment Division, Centre for Infectious Disease Prevention and Control. HIV/AIDS Among Aboriginal Peoples in Canada: A Continuing Concern*. Ottawa: Public Health Agency of Canada, 49-61.

Potts, K., 2006. "Music is the weapon: Music as an anti-colonial tool for Aboriginal people in Toronto." MA Thesis. University of Toronto.

Ricci, C., Flicker, S., Jalon, O., Jackson, R., Smillie-Adjarkwa, C., 2009. "HIV Prevention with Aboriginal Youth: A Global Scoping Review," *Canadian Journal of Aboriginal Community-Based HIV/AIDS Research*, 2, 1, 25-38.

Rose, T., 1994. *Black Noise: Rap Music and Black Culture in Contemporary America*. Hanover, New Hampshire: Wesleyan University Press.

Smylie, J., 2009. "The Health of Aboriginal Peoples," pp. 280-301. In Raphael, D., editor, *Social Determinants of Health*. Toronto, Ontario: Canadian Scholars Press.

Ali Lakhani, Vanessa Oliver, Jessica Yee, Randy Jackson & Sarah Flicker

APPENDIX A: TRANSCRIBED LYRICS

TORONTO WORKSHOP LYRICS — SONG TITLE: DO IT RIGHT

(VERSE 1)
I-I-I-I don't rhyme all the time
My shit isn't the best
Don't criticize 'cause you need to realize that you ain't familiar to this
It's a surprise in your eyes you might just get hypnotized
If you listen carefully the words I have spoken
You see I'm not jokin'
My heart is just broken
Gotta' keep on focusing
I open my eyes and look around me
All I can see is the trouble that surrounds me

(CHORUS X 2)
Every day is a struggle
You just gotta' hold on tight
Get up on your feet and do it right

(VERSE 2)
They can't understand me
My jeans may be baggy
That don't give the cops the right to grab me
Unhand me.
This world is so unhappy
Kids feel their life is so crappy
Lookin' at life like a suicidal trap b
It's like that see
And if you can't see
Then open your eyes and realize
That this world holds a big surprise
No matter where you are there will be lies
And people who criticize the way you look
They'll try and make you small
Try and get you shook
Make you wanna' retaliate with a left hoo-hoo-hoo-hook

(CHORUS X 2)

Don't criticize me. Don't isolate me.
'Cause we need to stay united
Accept me. Don't reject me.
'Cause we gotta' stand strongly
Keep the fire burning brightly
'Cause we all the same inside
(repeat)

215

KETTLE AND STONY POINT WORKSHOP LYRICS — SONG TITLE: THE PAIN

(INTRO)
I smoke like a chimney
I smoke like Bob Marley did
I used to get high like the birds and the planes
'Cause that's why we hope. For no more dope.

(VERSE 1)
The pain...
The pain
Naw it can't be explained
Puffin out my soul all I feel is the rain
The pain it feels so plain, I'll be sitting in my room and I'm going insane
Smoke too much Jane, You know it man. I got nothing more to gain.
Mary Jane is going to my brain smoke so much Jane, can't even be explained
I feel the pain — I'm going to go insane –the pain is going to blow my brain,
I'm in a plane —up in the sky because I'm so high. I'm so, so baked I need to eat some cake
My life feels like I'm on a train. Never ending I gotta use my brain. I gotta find an escapement to get out of this place. So I can live my life.

(CHORUS 1)
We need hope, we don't need no dope. *(YEAH)*
We need hope, we don't need no dope.
We need hope, we don't need no dope.*(YEAH)*
We need hope, we don't need no dope.*(YEAH)*
We need hope. We got hope. *(YEAH)*

My life — is a shitload of pain. There ain't no gain from the shit I do man
All I can do is smoke like a train I'm so high I feel like a plane
I ain't no bitch. I don't need to carry no gat or a vest. People see me in the hood and turn away and run like a bitch.
Sometimes we feel like we gotta' run away. But we know we gotta' stand and fight the pain. Show everyone how we run this place. First Nation people. It's like we got slapped. Straight across the face. See us. You know. We were here first.

(CHORUS 2)

(OUTRO)
Sometimes we feel like we gotta' run away. But we know we gotta' stand and fight the pain. Show everyone how we run this place. First Nation people. It's like we got slapped. Straight across the face. See us. You know. We were here first.

ANDERS LUND HANSEN[1]

Forty Years of System Change
Lessons from the free city of Christiania

we live in chilly times
everybody walks around and feels the chills
but we are doing ok here
here at Østre Gasværk

there's many, many dreams
that have gone down the drain
but I can still laugh
when I sit all alone —
all alone in my small shed

Kim Larsen, *Det er en kold tid (It is chilly times), my translation*
Lyrics from Christianiapladen (the Christiania record), 1976

Parallel to the climate change COP15 conference in Copenhagen in December 2009, the 'free city' of Christiania, Europe's largest and longest established squatter settlement, hosted a so-called climate bottom meeting which featured presentations on how "a number of sustainable cities and eco-village initiatives around the world meet the world's ecological, social, spiritual and economic challenges."[2] Many activists were drawn to Christiania during the climate summit to take part in the alternative two-week meeting and/or to take in the atmosphere of this unconventional urban setting. One of the slogans that was used in the streets of Copenhagen during COP15 was "System Change, Not Climate Change!" In this chapter, I shed a critical light on what Christiania has done, and is doing, to challenge not only the carbon economy but other conventions as well. My aim is to look at the obstacles that the community is facing and to provide an account of its promises as *one* element in a possible future "system change" (Harvey, 2010). I do so through a discussion of the different visions of the right to the city presented by mainstream society and the Christianites.[3]

Lessons from Christiania

Since 1971, Christiania has been a center of resistance, critique and a creative transformation of urban space. What started out as a squatter occupation of an abandoned military compound in central Copenhagen covering more than 85 acres, the equivalent of 10% of New York's Central Park, has developed into a home for almost 900 inhabitants.[4] Through continuous struggles, Christiania remains a laboratory for new modes of urban design, democracy and social and environmental justice. It is now a socialist/anarchist/liberalist urban social experiment (a success has many parents) that attracts tourists,[5] students, artists, architects and social scientists who come to experience and study this extraordinary urban setting.

What are the ingredients that have made Christiania into such a unique place? And how may other places be inspired by Christiania as an antidote to contemporary chilly social climates? An environmental — physical as well as social and physiological — awareness and responsibility has been an integral part of Christiania's value basis and urban praxis from the outset. Through continuously experimenting with ecological buildings, biological wastewater treatment systems, alternative energy, a 'car free city' politics, recycling stations, compost systems, the use of rainwater for flushing, and composting toilets, Christiania is seeking to reduce its ecological footprint on the world. In addition, a social responsibility system is in place in the form of several programs, including *Herfra og videre* (from here and beyond), which is a social support service operated in collaboration with

Copenhagen's municipal social services, an employment service, a health system, a health facility (*Sundhedshus*), and several other services relevant to solving complex social problems. Furthermore, 'culture' is seen as a cohesive force in Christiania, where different age, gender, ethnic and socio-economic groups are working and living side by side. The ideal is to develop a "feeling of belonging" for all groups, through, for example, jointly developed rituals and cultures around Christmas, funerals, baptisms, meetings, democratic practice and much more (Christiania 2007). Though far from perfect, the experiments with improving the environments in the free city have served as an inspiration for its surroundings throughout the years.

It would be nice if one could identify a 'Christiania doctrine' — a magical formula that could be used in the strife for a more democratic and just city. However, it is impossible to present a comprehensive account of Christiania's 40 years of insights into activism, alternative living and system change. I will instead share one moment in Christiania's history that I experienced when I lived in the community as a Christiania Researcher in Residence in May 2007. Since 2004, the locally supported and driven Christiania Researcher in Residence program offers residency for artists and academic researchers who are interested in generating important knowledge about Christiania (www.crir.net). The program has sponsored more than 40 projects on a variety of themes.[6] Visiting scholars and artists share their work and experiences through different modes of representation, including books, articles, photo, film, performance, and seminars. But before doing this, I will provide an account of some of the general obstacles that the community is facing.

Christiania — a contested space

Not surprisingly, Christiania is a prime target for the current "cultural battles" launched by a Danish right-wing government that came into power in 2001. The government's plan is to 'normalize' Christiania; the central objectives are to close down the cannabis market,[7] register and legitimize the building stock, and to abandon the principle of joint ownership of the land in favour of individual rental contracts and private property rights. A neoliberal revanchist strategy (Smith, 1996) stamped by the logic of a new urban imperialism (Lund Hansen, 2006), the design is to make way for gentrification, to harvest huge land rents (development gains) and displace the "economically unsustainable" (Copenhagen Municipality, 2005) population. Like in many other western cities, landscapes of urban slums produced by economic restructuring and disinvestment characterized the inner city areas of Copenhagen in the beginning of the 1970s. It was in this context that *Slumstormerne* (the slum troops) squatted on the former military compound on September 26th 1971, and since then the area has been used as a platform for the development of an alternative urbanism.

Since the establishment of Christiania, Copenhagen has generally experienced a huge transformation. The Danish government has not only strived to 'normalize' the free city of Christiania, but also to build a cross-border growth region together with southern Sweden to meet the global and local challenges of 21st century urban transformations:

> Copenhagen has one of the world's best business environments. ...The investment and business climate is world-class, combining low corporate taxation and a highly educated workforce with an international outlook and an outstanding quality of life. This is why Copenhagen is open for business. But Copenhagen is "Open" in many other ways too. Whether you are seeking cultural experiences, shopping, enjoying the city's quality of life or a great place to live, Copenhagen is open for you, which is reflected in the city's new brand: "cOPENhagen — open for you". (Copenhagen Capacity, 2010)[8]

The central actors on the urban political scene perceive Copenhagen as a node in the European urban system, and as a growth engine for all of Denmark. In this process the most powerful actors in the region have invested heavily in creating an identity for one whole region — the Oresund region — by linking Greater Copenhagen and the region of Scania in southern Sweden. Major investments include a motorway and railway bridge linking Copenhagen and Malmö in Sweden, an expansion of the international airport, a new subway line connecting the downtown with the airport, a new 'city tunnel' in Malmö facilitating train services between Scania and Copenhagen, and new major urban development projects such as Ørestaden, Holmen (next to Christiania) and Havnestaden. Other material manifestations include symbolic works of architecture, such as the Turning Torso in Malmö, Arken (The Ark), the new museum of modern art, and Den Sorte Diamanten (The Black Diamond), the new waterfront annex to the Royal Library, a new concert hall in Ørestaden, and a new opera house on the harbourfront in Copenhagen. The opera house is a 'gift' to the city from Mærsk Mc-Kinney Møller, the owner of a major multinational (shipping, oil, airline etc.) corporation, and the most powerful capitalist in Denmark. As a powerful actor in what Cindi Katz (2001) calls the age of vagabond capitalism, Mærsk Mc-Kinney Møller offers gifts rather than pays taxes. The gift offers convenience because the donor decides what to give — and it is not polite to complain about a gift.

Other material manifestations of the 'new economy' include the new built environments for the main actors (the information technology, and finance, insurance and real estate sectors), including luxury hotels, restaurants, conference centres and shopping malls, such as Fisketorvet on the harbourfront, and luxury housing and publicly financed renewal of inner city housing to attract the 'new middle class', the employees of the 'new economy'. These processes of gentrification, generated by public policy, entail the displacement of inner city residents who do not fit in the 'new creative economy' and Copenhagen's "world-class business climate" aspirations (Lund Hansen et al., 2001; Larsen and Lund Hansen, 2008).

The city is thus open to some people while closed to others. In light of this changing urban scene, Christiania is under considerable pressure. The neoliberal urban strategies behind the production of the "New Copenhagen" (Bisgaard, 2000) is applied to Christiania through the discourse of 'normalization', that is, the 'legitimization' of its building stock and the 'privatization' of its common lands. But the potential 'tragedy of the commons' of Christiania is not happening without a struggle. These struggles do not only take the form of violent street battles, but is also fought at a more subtle policy level. As one Christianite states: "It [the government] is grinding us down slowly. They realize that using bulldozers is not a good idea. Bureaucrats are good though: they work! And suddenly it [Christiania] becomes a 'nice' area — and damn boring. I can't stand niceness!" (Guldbrandsen, 2005).

As a possible counter strategy to the government's gentrification strategy, a collaboration between Christiania and KAB (a non-profit rental housing association) is being established (KAB, 2004). The idea is to transform Christiania into an independent non-profit rental housing association and a foundation for small businesses. The future will show if a marriage of the special forms for anarchism we find in Christiania and the reformed socialist practice of KAB is a viable solution for Christiania. Co-optation and misrepresentation constitute key challenges in this context. Internal turf wars, reflecting the wide spectrum of income, age, gender and ethnic diversity that is a main ingredient for Christiania — and other communities who are fighting for their right to the city, could potentially divide and destroy the community (Angotti, 2008). On the other hand, the roots of Christiania's struggles can be compared to tenant struggles against urban renewal and gentrification as well as the environmental justice movement's struggles against suppression of rights to the commons. Seen through the lens of David Harvey's (2003) concept 'accumulation by dispossession', these struggles resonate very well with many of the struggles that form the agenda of participants in alternative globalization movements.

Christiania's struggles for the right to the city are multi-scalar and multi-facetted. The current main strategy is a court case against the Danish state, where Christiania claims squatter's rights. There is also a long tradition of local politics and art practices. Best known are the actions created by the theater group *Solvognen* (the Chariot of the Sun) from 1969 to 1982, and 2006 to the present. Some famous events include the invasion of the Native Americans at *Rebilfesten* (the celebration of Danish-U.S. relations) in July 1976 and the Guantanomo happening in July 2006.[9] Christiania is also notorious for its everyday struggles and activism. In the following section, I will share one moment in Christiania's long history of struggles that I experienced as a Researcher in Residence. Early Monday morning, March 14, 2007, the police arrived and demolished *Cigarkassen* (The Cigar Box) (a small one-family house, or small shed), located on the ramparts in an area named Midtdyssen in Christiania, only to see it rebuilt the next morning. According to the Danish state's normalization plan, the building was an 'illegal structure' with 'illegal inhabitants' (homeless people). The demolition was preceded by street battles between activists and the police. The state took action despite the fact that a court decision on these issues would be announced a few days later. Five days after the events a decision was announced — in favor of Christiania.[10]

Wanted: political goodwill and commitment!

On Tuesday morning, March 15, 2007, *Cigarkassen* reappeared as a phoenix from the ashes, less than a day after the police had demolished the building. "Come again!" read a three-meter long red banner that was lashed to a tree. The house was approximately four by two meters in length and breadth and three meters high. It was constructed of wood pallets, laths and veneer sheets. It was a solid construction that even had a small annex by the water with the inscription: "The reconstruction team strikes again." The idea of the annex was to signal that "when they tear down a house — we are building two houses," explained a young woman. The main house had an awning that said "Welcome". The sign was made "in honor of the helicopters" explained the activist who painted the message. Inside the house all was neat and well lit. There was a fixed bench and a picture hung on a nail on the wall. Most striking was the 'fire place' with painted hot flames creating warmth in the otherwise chilly time we live in.[11]

Figure 1: Cigarkassen's new 'fire place', with its painted hot flames – creating warmth in the otherwise chilly times we live in (photo by author).

The evening before, 25 to 30 activists — most of them in their twenties — completed the action. They worked throughout the night with great enthusiasm. Meanwhile, street battles in the neighboring area of Christianshavn between the police and supporting activist groups occurred. Many activists were arrested, among them my neighbour's son. I was drawn towards the place that started the day's events. I offer the following account of the events of the night to give an impression of the people, their efforts, and convictions as they rebuilt *Cigarkassen*:

> The reconstruction is well in progress when I arrive. Tools are changing hands and building materials are being brought into use. The warm May evening is full of a positive energy that stands in stark contrasts to the street battles fought a few miles away. As darkness falls, power cables and halogen lamps are retrieved and work continues. "It was damn good, we were here early," says a man around thirty years of age to his friend while opening a Tuborg beer. Both are dressed in white carpenter's pants and they confirm that they are professional craftsmen. They report: "We said very quickly that such and such a small solid house could be constructed. Everybody accepted the plan and now you can see that it works." Both men are satisfied with their efforts. People at the building site talk about the government, police violence and the normalization of Christiania.

Figure 2: Activists are working eagerly on the rebuilding of Cigarkassen (photos by author).

"Does anyone here live in Christiania?" asks a young girl who is helping out coordinating the construction work. "Yes, here", says a man in his 40s. "We need some building materials. Do you know where we can find something we can use?" It is being coordinated and handled. Later, I am told by a Christianite who participates in the work that several activists are previous Børne Hus (Christiania's daycare) children.[12] Water, cola, beer, coffee, tea and sweets are being fetched. Later, there is someone who tries to get some food for the whole group: "Is there anyone who'd give their number out, so I can call and see if you are still here? It may take some time to get the food." Two or three activists volunteer their number. One suggests getting a hold of a joint. But this proposal is refused, the group feeling that "a high activist is a slow activist."

The atmosphere is good, though people are aware of the situation's seriousness. "Just imagine, maybe we will get arrested? Do you think this is illegal?" a young woman wonders. People talk about Christiania's building stop and the implications of the action. An activist approaches the sight: "Does someone want to replace one of the guards who has been sitting down the road and kept an eye on the police for some hours now? One of them should like to be replaced." The group has a clear awareness of the event's historic potential, and it is clear that taking part in the evening/night's events is a considerable action to add to any activist's CV. References are made to Byggeren, a similar action in Nørrebro (the Northern part of the city) in the 1970's and 1980's and people take pictures and films.[13]

Figure 2: *Cigarkassen* — rebuilt, May 15th 2007 (photo by author).

The next morning, the building was finished. The morning highlighted the imaginative colourful decorations and unambiguous words, which effectively expressed what the night's action had been all about: "Wanted: political goodwill and support!"

During the day, Christiania's press office made sure that the nationwide media got wind of the story. A significant number of reports covered the rebuilding effort, but the vast majority of the headlines focused on the street battles, barricades and burning cars in the surrounding neighborhood, Christianshavn. In isolation, the rebuilding of a small wooden house may not seem significant. But the action could be regarded as an active politics of scale — an important *symbol* of Christiania's proactive fight for its 'right to the city'.

The right to the city: towards "system change"

The dream of a market utopianism persists despite the recent economic crises. In *A Brief History of Neoliberalism*, David Harvey (2005) suggests that in the event of a conflict between the health and well-being of financial institutions and people, financial institutions will win every time. He also proposes that under neoliberalism government resources are primarily used to create a good 'business climate'. Throughout the world, shrinking governmental resources are increasingly redirected towards the support of business' needs at the expense of social budgets — often imposed through a 'chock doctrine' (Klein, 2007). Neil Smith has stated that neoliberalism is dead, dominant and defeatable but reminds us that a dead rattlesnake "can still strike, and neoliberalism, however dead, remains dominant" (Smith, 2008: 1-3). But is it defeateable? And can we learn from alternative urban communities like Christiania?

The French philosopher Henri Lefebvre (1991) saw the emancipatory potential associated with the multi-scalar processes behind the production of space. Lefebvre (1996) argued that the processes of urban transformation offer opportunities for marginalized social groups to claim 'the right to the city'. Christiania is an excellent example of such a struggle. In this chapter I have shed a critical light on two very different visions of the right to the city. First of all, private property is the dominant right to the city throughout the history of capitalism and it has been at the core of the 'neoliberal revolution'. Cities throughout the globe have become important spaces of neoliberalism and an entrepreneurial (Harvey, 1989) urban politics accommodating investors, developers and the so-called creative class (Peck, 2005). Those in charge of city governance use Margaret Thatcher's TINA acronym (There Is No Alternative) to support neoliberal policies and make them the norm in a post-political city (Swyngedouw, 2007). And this is also the case in "cOPENhagen" where

Christiania is fighting *against* the normalization of private property and *for* a very different right to the city, that is, a collective
right to land and housing. Through the example of the rebuilding
of Cigarkassen, I have presented one moment in a long history of
struggles. I suggest that collective activism, dedication, improvi-
sation, art, humor and practicing a politics of scale are important
elements in Christiania's 40 year struggle for the right to the city
— and hence may still be important elements in a future "system
change" in chilly times.

* * *

Anders Lund Hansen *is an Assistant Professor of Human Geography at
Lund University. His research is on the political economy of space,
uneven development, urban social geography, gentrification, space
wars, and, more recently, East Asian urbanism. Since 2007, he has
served on the steering committee of the Christiania Researchers In
Residence program (www.crir.net).*

ENDNOTES

[1] anders.lund_hansen@keg.lu.se. The financial support from the Jan
Wallander and Tom Hedelius Foundation (research grant number W2007-
0055:1) is gratefully acknowledged.

[2] The meeting took place the 5th–18th of December 2009. It was organized
by a coalition consisting of Christiania's Culture Association, The Network
for the Conservation of Christiania as a Green Urban-biotope, two Agenda
21 centers in Copenhagen (Sundby and Inner City), LØS, The Danish
Association for Eco-villages, and GEN, Global Eco-village Network — in
collaboration with many individuals, associations and groups.
www.climatebottom.dk/en, accessed 15 June 2010.

[3] See: http://www.righttothecity.org.

[4] According to Copenhagen Municipality's Statistical Office, 878 people
(hereof 167 children) were registered inhabitants of Christiania on
January 1, 2003. Demographically, middle-aged couples dominate
Christiania. The average income is 106,000 DDK, which is almost half of
the average income in Copenhagen in general. 33% is connected to the job
market (56% in Copenhagen in general).

[5] Together with Tivoli and Den Lille Havfrue (The little Mermaid),
Christiania is one of Copenhagen's main tourist attractions.

[6] Here is a selection of the CRIR projects: self-government and self-policing; a comic strip, "what is the mystery", published in Ugespejlet (the local newspaper); the repair of a mural painting; social perspectives on new housing areas; video art; a search for "the perfect location"; a documentary film titled "Our heart is in your hands"; a dance performance named "Half machine"; gaps in the political economy of Christiania's land — Christiania as a space of hope; the practice of 'ownership' in Christiania; the relations Christianites have to the pictures often shown of Christiania in the media; gentrification; the dogs of Christiania; space wars; lessons from Berlin; how to rebel in a society that is already in a state of rebellion?; democracy; social justice; surveillance of government buildings, symbolically redressing the balance; "You film us, we film you"; a quest for Utopia; Christiania's symbolism; paths through utopias; paradise; a location-sensitive model for a future history museum on Christiania; GNH — Gross National Happiness; sleeping in spandrels; high on life; movie on the residents' relation to the rest of Copenhagen; "Bevar Christiania"; conceptualizing the 'ecosocial'; creative Copenhagen; community experiments in collaborative homes and lifestyles; public space.

[7] In the spring of 2004, without any major violent coordinated resistance from the inhabitants of Christiania, the police closed down Europe's largest open air cannabis market on "Pusher Street" where 'pushers' sold their products from 30 booths in the center of the community.

[8] Copenhagen Capacity is Copenhagen's booster organization. http://www.copcap.com/content/us/doing_business/why_copenhagen/copenhagen_-_open_for_business.

[9] See: http://www.vestfilm.dk/christiania/solvognen/chariotofthesun.html.

[10] Another version of this 'report' was published in Danish in *Ugespejlet* (Christiania's newspaper) and the neighborhood newspaper *Christianshavneren*, June issue, 5, p.4. (http://www.christianshavneren.dk/upl/8962/2007juniside113.pdf).

[11] "The chilly times we live in" is also a reference to the popular tune 'It is chilly times' by Kim Larsen, who is a very popular folk singer in Denmark. The tune was played loudly throughout the center of Christiania during these days in May 2007. The Danish Social Democratic Prime Minister, Anker Jørgensen, did also use the reference in one of his speeches at the end of the 1970's to characterize the global and national economic crises.

[12] The Børne Hus is one of Christiania's own 'institutions'.

[13] Some of the events of the day can be seen at: http://www.youtube.com/watch?v=D1qxIEYwMOU.

REFERENCES

Angotti, Tom, 2008. *New York for Sale: Community Planning Confronts Global Real Estate.* Cambridge MA: MIT Press.

Bisgaard, Holger, 2000. Interview with Holger Bisgaard, Head of Planning for Copenhagen Municipality, Copenhagen, 26 January.

Christiania, 2007. Christiania som agenda 21 område - et exempel til efterfølgelse (Christiania as Agenda 21 area - an example to follow), Christiania. http://www.christiania.org/inc/pdf/07_tekst/07_ca_agenda21_folder.pdf, accessed 1 June.

Copenhagen Capacity, 2010. http://www.copcap.com/content/us/doing_business/why_copenhagen/copenh agen_-_open_for_business, accessed 1 June.

Copenhagen Municipality, 2005. *Boliger for alle — boligplan for Københavns Kommune 2005-2008.* Forslag. Copenhagen: Københavns Kommune.

Guldbrandsen, 2005. Kampen om staden (The battle of Christiania), Television Documentary, DR1, 5 January, 8 pm.

Harvey, David, 1989. "From managerialism to entrepreneurialism: the transformation in urban governance in late capitalism," *Geografiska Annaler*, 71B, 3-17.

Harvey, David, 2003. *The New Imperialism.* Oxford: Oxford University Press.

Harvey, David, 2005. *A Brief History of Neoliberalism.* Oxford: Oxford University Press.

Harvey, David, 2010. *The Enigma of Capital, And the Crises of Capitalism.* London: Profile Books.

Klein, Naomi, 2007. *Shock Doctrine: The Rise of Disaster Capitalism.* London: Penguin.

Lund Hansen, Anders, 2006. *Space wars and the new urban imperialism.* Lund: Department of Human Geography, Lund University.

Lund Hansen, Anders; Andersen, Hans Thor & Clark, Eric, 2001. "Creative Copenhagen: Globalization, urban governance and social change," *European Planning Studies*, 9, 7, 851-869.

KAB, 2004. En mulig vej for Christiania – som fond og almen boligorganisation (A possible way for Christiania – as a foundation and a non-profit housing association), Copenhagen, KAB Bygge- og Boligadministration. http://www.kab-bolig.dk, accessed 1 June 2010.

Katz, Cindi, 2001. "Vagabond capitalism and the necessity of social reproduction," *Antipode*, 33, 4, 708-727.

Larsen, Henrik Gutzon & Lund Hansen, Anders, 2008. Gentrification - gentle or traumatic? Urban renewal policies and socio-cultural transformations in Copenhagen, *Urban Studies*, 45, 2429-2448.

Lefebvre, Henri, 1991. *The production of space*. Oxford: Blackwell.

Lefebvre, Henri, 1996. The right to the city. In E. Kofman and E. Lebas (trans. and eds.) *Writings on the city*. Berkeley, University of California Press.

Peck, Jamie, 2005. "Struggling with the Creative Class," *International Journal of Urban and Regional Research*, 29, 740-770.

Smith, Neil, 1996. *The New Urban Frontier: Gentrification and the Revanchist City*. New York: Routledge.

Smith, Neil, 2008. *Neoliberalism is Dead, Dominant and Defeatable – Then What? Human Geography*, 2, 1-3.

Swyngedouw, Erik, 2007. "The post-political city," pp.58-76. In G. Boie and M. Pauwels, editors, *Urban Politics Now: Re-imagining Democracy in the Neo-liberal City*. Amsterdam: NAi Publishers.

ADRINA BARDEKJIAN AMBROSII

Marginal Medleys
How Transition Towns and Climate Camps are relocalizing the global climate crisis

If you have built castles in the air, your work need not be lost; that is where they should be. Now put foundations under them. — **Henry David Thoreau**

When you first hear the terms Transition Towns and Climate Camps, what comes to mind? Romanticized images of twenty-something bohemians and hippies? Long haired girls in patchwork dresses picking fruits and flowers from trees? Young men wearing t-shirts with political icons while picketing or shouting protests? Children running barefoot through a maze of pitched tents with the expanding sky overhead as the roof of their existence?

Much like in the fictional world created by Ernest Callenbach in his novel, *Ecotopia* (1975), the Transition movement is concerned with peak oil, climate change and economic impacts; unlike the novel, these societies are very real. Since 2006, Transition Towns and Climate Camps have emerged like guerilla groups in response to a consumer-based, carbon-dependent society. But there are some differences between them. Started by Rob Hopkins, a professor of permaculture in the UK, the Transition Town initiative is a long-term reformist strategy that encourages communities to adopt Energy Descent Action Plans.

In contrast, Climate Camps are temporary campaign gatherings to denounce the destruction of the climate, advocate for a zero-carbon culture, and work for an alternative society.

Once thought of as the catalysts for creating a local planet; globalization and the capitalist economy have fragmented and impoverished local economies that were once self-sustaining; they have inadvertently sought to break the traditional familial impulses to collaborate and cooperate with one another and fostered instead a sad state of competition, isolation and pressure (Carlsson, 2008). Arguably, the Western lifestyle has left people empty and grasping for a lost meaning in life (Brook, 2009) and questioning how to define prosperity and success.

Global warming, more than war or political upheaval, stands to displace many millions of people. And climate change is being driven by the fossil-fuel-intensive lifestyles that we enjoy so much.
— New Economics Foundation (2004)

The very real problems of peak oil and climate change overlap to paint a dismal portrait for a carbon-dependent future. People in developed countries consume on average four barrels of oil for every new one discovered and, between 2011 and 2015, supplies will begin to diminish (Hopkins 2010). The era of cheap oil has ended and things are about to get very expensive. But what if those people, in other words, we, did not depend on oil as excessively as we do now? What if we began to take steps to move away from this lifestyle? Really move away. We are at a crossroads and true to its name, the Transition Town movement claims to steer the way past the tipping point towards a new and important opportunity of practical, plausible, relocalized living.

The Transition Town movement is creatively strategizing solutions for a healthier, sustainable and just future, without the toxic, armageddon-style messaging that usually accompanies environmental advocacy (Hopkins, 2010; Brook, 2009). This is very alluring as the movement acknowledges the precarious state of the natural environment but claims to provide practical solutions through an iterative process which is constantly changing according to group, country and culture. It is self-organizing and devoid of a top-down politically-driven approach. Communities plan their own process and do what works for them based on their existing resources, skills and assets (Brook, 2009).

The movement has three objectives: 1) planning and designing initiatives to move away from carbon-dependency; 2) Powering Down by rebuilding the local economy to support essential requirements; 3) Powering Up by putting in place a renewable energy infrastructure that can actually sustain a non-consumer-based society. The main focus of the 12-step model outlined in the *Transition Town Handbook* (Hopkins, 2008a) is looking at the practicalities of relocalization and advocating both their benefits and importance.

Emphasis is placed on local ecological resilience (Brook, 2009). The Western love affair with urbanity and incessant praise of globalization has made for vulnerability in the face of environmental catastrophe. Be it individually or collectively (that is, in a city), people cannot take care of themselves for the most basic necessities, such as finding food without a grocery store or creating a fire for heating or cooking. It is at the community or neighbourhood level where unity can be obtained through skill-sharing and adaptable ingenuity. That is not to say that the Transition model cannot be applied in cities; by working at a neighbourhood scale, then creating city-wide networks, the transition movement claims to be quite an effective matrix (Hopkins 2010). Currently, there are 30 different initiatives across London and recently there was the first meeting of "Transition London" as a concept and network.

A major attraction is that the Transition movement is very much in the "here and now" for people willing to jump into action immediately. Yet, despite a radical desire to bypass the long and sticky political red tape, a program of intentional relocalization will only be thoroughly successful if it is driven by political vision and leadership from all levels of government, regardless of how long that may take. For example, Bristol was the first Transition City working and engaging with city counsel and as a result was instrumental in passing a peak oil resolution. "We need communities making the unelectable, electable and being the real drive and the push for making this happen" (Hopkins, 2010).

There are currently 300 formal Transition projects around the world (UK, Ireland, Canada, Australia, New Zealand, U.S., Italy and Chile) and several thousand smaller groups who are still determining whether to become official projects. The last action in the 12-step transition model is to create an Energy Descent Action Plan. The UK town of Totnes, also the first UK Transition initiative, has recently published their Action Plan, *Transition in Action: Totnes and District 2030*. Specifically, some activities described in realizing their society transformation include: setting up local companies where individuals in the community can purchase shares; creating a local currency; strategic planting projects, such as hybrid walnut trees; distributing seed boxes; sponsoring "sustainability week" events and education exhibits; community cafes; and developing a garden-share program that matches people who want to garden with people who have land that they are not using.

Though these activities may seem *ad hoc*, the Transition movement sees them as conjoining parts in the overall context of intentional relocalization of communities. These initiatives are intended to foster a culture of stewardship by moving people to action and offering them applied, practical and realistic designs by which they can live (Hopkins 2010). Communities are coming together across the world through creative and adaptable ways of dealing with these adverse results and rebuilding local economies as their core objective. We are headed back, towards a style of living where relocalization, not technology, is the sustainable solution.

Climate is an ill-tempered beast, and we are poking it with sticks.
— Mark Maslin author of Global Warming (2004).

Having been initiated in the UK, on a less permanent scale, Climate Camps have emerged in various countries (ie. Canada, France, Netherlands, Scotland, Australia, Iceland). Their message is direct and clear: we must become a zero-carbon culture. Aggressively campaigning against the destruction of our climate, these activists protest the "business as usual" model (Cunningham, 2008). Unlike Transition Towns, their more permanent counterpart, Climate Camps promote direct action and sustainable living (Beck, 2008).

To date, there have been numerous camps targeting specific causes and hosting workshops on a variety of issues related to climate change and social justice such as coal mining, class composition and economic oppression. Examples in the UK include: protesting against a proposed new runway at Heathrow airport (2007); protesting against a proposed new coal-fired power station, Kingsnorth Power Station (2008); and Camp in the City, a protest associated with the G20 London Summit (2009). There were also six climate action camps organized across Canada in 2009 by Greenpeace with the intention of teaching "peaceful civil disobedience" (Diotte, 2009). Camps are free to attend and are run by consensus-decision making (Heimbuch, 2010). They also operate in an eco-friendly manner; they may use solar panels and wind turbines for power, pedal-powered laundry, biodiesel from recycled cooking oil for vehicles, vegan food and compost toilets (Handout, 2007). An example of a current camp is the Camp for Climate Action 2010 which is targeting the Bank of Scotland.

Though Climate Camps are temporary campaign gatherings, there is a tendency to look at people in a collective group and miss the individual contributions. Photographers have captured intimate portraits of climate activists in their camps (Sherratt, 2008). These images allow us to reevaluate our stereotype of climate campers; they are all ages and from diverse backgrounds. "The Climate Camp is a place for anyone who wants to take action on climate change; for anyone who's fed up with empty government rhetoric and corporate spin; for anyone who's worried that the small steps they're taking aren't enough to match the scale of the problem; and for anyone who's worried about our future and wants to do something about it" (Heimbuch, 2010).

The fishermen know that the sea is dangerous and the storm terrible, but they have never found these dangers sufficient reason for remaining ashore.

— Vincent van Gogh

The environmental movement has often been criticized for being too white and exclusive (Beck, 2008; Jones, 2007; Gosine, 2005; Harter, 2004). As such, responses to Transition Towns and Climate Camps vary according to interest group. However, the main criticism of the Camps is that it is another anti-capitalist campaign available as a luxury solely to the middle-class. It has also been noted that the Camps have been referred to as "alternative lifestyle festivals" (Cunningham, 2008) and the Camp parley has been criticized as having little concern for class analysis (Douglass, 2008). It also raises a major question: Is it realistic for people to break away from their daily lives to join a Climate Camp? Would they take vacation time from work to do this?

Of the Transition initiative, critical adjectives used against them have included, "dangerous", "rubbish" and "inefficient" (Hopkins, 2008b). An interesting spin is the notion that neither the Climate Camps nor the Transition movement are enough and are indicative of the too-little-too-late attitude. For example, "In an evening debate on the role of the state, environmental writer George Monbiot expressed his concern that voluntary, low-carbon lifestyle changes aren't sufficient to deal with the urgency of the crisis and that government intervention is need-ed" (Beck, 2008). However, considering that one of the core aspects of the Transition movement is to get various levels of government on board, the integration of communities such as Transition Towns and Climate Camps with their goal of govern-mental support and intervention could be an important oppor-tunity (Hopkins, 2010; Douglass, 2008; Shield, 2008).

The desire by activists to acknowledge the environmental crises (both human and ecological) as an opportunity is palpable. "The next 10 years will see massive changes across all sectors of society. We believe that this is a massive opportunity to create a society that values people for more than their ability to create profit. It is a massive opportunity to re-invest in communities and local economies and to create thriving, healthy environ-ments. But it will not come without a struggle and it is an oppor-tunity we will miss if we, the huge global majority who stand to lose or gain over the coming decades, do not work together" (Shield, 2008).

In the depth of winter I finally learned that there was in me invincible summer.

— Albert Camus

Against the backdrop of environmentalism, skill-building, and collaboration the revolving themes of these movements are nar-rative, positivism and cultural inclusivity (Hopkins, 2010;

Heimbuch, 2010; Brook, 2009). In the end it comes down to a philosophy of living.

Climate change is not solely about melting polar ice caps, ocean acidification and droughts; it is about the people, the communities, and the families living through those events. One of the most important aspects of Transition Towns and Climate Camps are the cultural stories that are brought to, or that evolve from, the movement itself; it is the various narratives that drive it. The need for new cultural and intimate stories is integral to its success. In essence the individual voices are not only heard, but are welcomed. As such, it pointedly brings people together at the local level to achieve a common goal in the face of energy constraints, a very real and recognized problem. Whether or not the movement is a middle-class, anti-capitalism campaign does not change that the intent behind it is a sincere concern for the natural environment and our future within it (Hopkins, 2010; Brook, 2009; Carlsson, 2008; Beck, 2008; Cunningham, 2008).

As mentioned in the introduction to this book, climate justice advocates argued in Copenhagen and Cochabamba for system change; Transition Towns and Climate Camps are two of those incremental steps to get there. We must not only be politically and economically aware, but environmentally and socially aware as well. The time for climate change has passed; the time for system change and climate justice is now.

* * *

Adrina Bardekjian Ambrosii is a doctoral candidate in the Faculty of Environmental Studies at York University.

REFERENCES

Beck, Juliette, 2008. Social Change not Climate Change: UK Camp for Climate Action. http://www.yesmagazine.org/issues/columns/social-change-not-climate-change, accessed 10 June 2010.

Brook, Isis, 2009. Turning up the heat on climate change: Are transition towns an answer? *Environmental Values*, 18, 2, editorial.

Callenbach, Ernest, 1975. *Ecotopia*. New York, New York: Bantam Books.

Carlsson, Chris, 2008. *Nowtopia: How Pirate Programmers, Outlaw Bicyclists, and Vacant-Lot Gardeners Are Inventing the Future Today!* Oakland, California: AK Press.

Cunningham, John, 2009. A climatic disorder? Class, coal and climate change. http://www.metamute.org/en/content/a_climatic_disorder_class_and_climat e_change_in_newcastle, accessed 9 June 2010.

Diotte, Kerry, 2009. Greenpeace camp for 'climate defenders': Hundreds of people being trained in civil disobedience. *The Edmonton Sun*. (August 29th August 2009, 4:20am), http://www.edmontonsun.com/news/alberta/2009/08/29/10664046-sun.html.

Douglass, Dave, 2008. Climate Camp report, http://www.indymedia.org.uk/en/2008/08/407011.html, accessed 9 June 2010.

Gosine, Andil, 2003. "Myths of Diversity," *Alternatives: Canadian Environmental Ideas and Action*, "Colours of Green" issue, 29, 1 (Winter), 12-17.

Handout. 2007, Climate Change, Capitalism & the Camp for Climate Action. http://www.climateactionnetwork.org.uk/toolkit/outreach_ideas/materials_f or_talks_and_workshops/handout_for_climatecamp_workshop.pdf, accessed 9 June 2010.

Harter, John-Henry, 2004. "Environmental Justice For Whom? Class, New Social Movements, and the Environment: A Case Study of Greenpeace Canada, 1971-2000," *Labour/Le Travail*, 54 (Fall), 83-119.

Heimbuch, Jaymi, 2010. What's a Climate Camp? See How Activists on the Job Live. Photo slideshow. Photo. Mark Sherratt. http://www.treehugger.com/galleries/2010/04/what-is-a-climate-camp-see-how-activists-on-the-job-live.php, accessed 9 June 2010.

Hopkins, Rob, 2010. *Rob Hopkins: Climate Change, Peak Oil and Transition Towns*. Helsinki. http://vimeo.com/11832667, accessed 11 June 2010).

————, 2008a. *The Transition Handbook: From oil dependency to local resilience*. Green Books.

————, 2008b. Responding to various critiques of Transition. http://transitionculture.org/2008/09/05/wading-through-various-critiques-of-transition/, accessed 9 June 2010.

Jones, Van, 2007. Eco-Apartheid: Why is the green movement so lily white? *Common Ground*. (April). http://oilsandstruth.org/eco-apartheid-why-green-movement-so-lily-white, accessed 9 June 2010.

Shield, Peter, 2008. Just Transition — Call from the 2008 Climate Camp. http://www.naturalchoices.co.uk/Just-Tranisition-Call-from-the?id_mot=10, accessed 9 June 2010.

Sturgeon, Noël, 2009. *Environmentalism and popular culture: Gender, race, sexuality and the politics of the natural.* Tucson: University of Arizona.

Wilson, Alexander, 1992. *The culture of nature: North American landscapes from Disney to the Exxon Valdez.* Cambridge, MA: Blackwell.

DEBORAH BARNDT

Dig Where You Stand!
Food research/education rooted in place, politics, passion, and praxis

A popular education program in Sweden, called "Dig Where You Stand" (Lindqvist, 1979), encourages a 'bare-foot' research and educational process that begins with ourselves and our immediate environment to uncover the history beneath our feet and critically engage with it. It is also an apt mantra for The FoodShed Project, a new collaborative research project involving over 30 local food initiatives, in collaboration with faculty and students at the University of Toronto and York University.[1] In fact, our focus on the southern Ontario foodshed aims to recover the history and strengthen a burgeoning innovative network of agro-ecological farmers, farm educators, land stewards, urban agriculturalists, environmental and community food security organizations, ethnic culinary initiatives, certifying bodies, producer and consumer coops, public agencies, food educators and researchers.

The framing of the project around the concept of the foodshed counters the geopolitical thinking that has shaped the global food system over the past 60 years. Since the turn of the millennium, there has been ample evidence that this model of corporate industrialized and export-oriented agriculture coupled with neoliberal trade policies has been a major contributor to climate change. Besides its inordinate dependence on vast quantities of water for monocultural agroindustrial production, it produces greenhouse gases (carbon dioxide, methane and nitrous oxide) and depends on fossil fuels for long-distance transport. Many aspects of the production, distribution, and consumption practices of the corporate and chemicalized agrifood system threaten biodiversity, human and ecological health (Barndt, 2007; Roberts, 2009).

The FoodShed Project is being shaped by its collaborators who are part of a broader movement for climate justice, guided by principles of equity, sustainability and civic engagement. The research and educational process we are crafting is rooted in place, politics, passion and praxis, integrating processes of decolonization, popular education, community arts and participatory action research. This essay outlines each of these interrelated methodologies and illustrates them with examples from FoodShed Project partners.

Rooted in place: decolonization

To be rooted in place, we need to be conscious of both the historical and the geographic context that shapes who and where we are now. The monument of Christopher Columbus with an Indigenous woman at his feet, erected in Peru in 1876, could be located anywhere in the hemisphere or even in Europe (Figure 1).

It immortalizes the white male European 'discoverers' who brought 'civilization'— epitomized by opulent clothing, a cross and an upward gaze — to the 'savages/heathens,' here a naked Indian maiden. Even though this artistic representation is more than 150 years old, people today inevitably still see their own lives within the persistent and intersecting power relations represented here: sexism, heterosexism, classism, militarism, religious evangelization, racism.

Figure 1: Columbus with Indigenous woman at his feet (photo by author).

How does the Columbus statue reflect your history and cur-

Figure 2: New Canadian farmers at Farmstart's McVean farm near Brampton (photo by author).

rent relationships? How are similar relations of power operating in the global food system? How can we decolonize the current debate about climate change, in particular, revealing the colonial dimensions of the global food system?

Decolonization can be seen as comprising several different processes: acknowledging the history of colonialism; working to undo the effects of colonialism; striving to unlearn habits, attitudes, and behaviours that continue to perpetuate colonialism; and challenging and transforming institutional manifestations of colonialism. Indigenous and post-colonial theorists have helped us unpack colonial notions of knowledge and knowledge production while also probing the ways that colonized peoples are speaking back from the margins, reclaiming not only their land but also diverse ways of knowing and communicating (Tuhiwai Smith, 1999; Loomba, 1998; Spivak, 1988). How can we ensure that these voices are heard in the struggle for climate justice and that we all learn from other ways of envisioning the Earth and our relationship to all living things? How can we draw upon the knowledges of diasporic populations to develop food alternatives that are place-based but not place-bound?

York University is located on property that was originally filled with corn that fed 1,000 inhabitants of a Seneca village.

Decolonizing would involve learning about the history and practices of Aboriginals native to this land, as well as drawing upon the rich diversity of practices brought to this land by the diasporic populations who are increasingly a majority (and may have Indigenous roots in another country). Two FoodShed partners have created programs that honour the agricultural knowledges and food traditions of new Canadians, for example. For a nominal cost, FarmStart (www.farmstart.ca) offers plots of land on the outskirts of suburban Brampton to new Canadians so that they can test their own seeds in a new ecological context, and incubate new farms (Figure 2). In the heart of Toronto, the urban agriculture coordinator at The Stop Community Food Centre (www.thestop.org), an African agronomist is experimenting in growing tropical fruit in a spacious greenhouse at the Wychwood Barns.

Rooted in politics: popular education

(The) general liberal consensus that 'true' knowledge is fundamentally non-political (and conversely, that overtly political knowledge is not 'true' knowledge) obscures the highly organized political circumstances obtaining when knowledge is produced.
(Said, in Young, 2003: 59)

The implicitly political nature of education, alluded to above by Edward Said, was profoundly articulated by Brazilian educator Paulo Freire, whose seminal book *Pedagogy of the Oppressed* has influenced popular educational projects and social movements all over the world. Freire contended that education is not neutral, that it must start with the experiences and perceptions of the learners, that the content should be drawn from their daily lives, that the teachers are learners and learners teachers. His problem-posing approach to education encouraged learners to name the social contradictions they faced and to consider how they might act collectively to transform them. This process of 'conscientization' integrated the personal and political, the individual and collective, action and reflection.

Freire built on an analysis of power developed by Antonio Gramsci (1971). The Italian Communist journalist introduced the concepts of hegemony and counter-hegemony as a way of understanding power and struggle for change. His notion of hegemony is dynamic, framing power as relational, or persuasion from above as well as consent from below. Gramsci suggested that dominant groups maintain ideological control through intellectual and moral persuasion, winning the hearts and minds of people who might not even share their interests. Struggles for power by marginalized groups represent 'counter-hegemonic' forces that challenge and transform this dominant hegemony. People must consent to dominant ideas and practices in order for hegemony to work. Ideological institutions such as schools, media, and advertising are critical to this process, and so any efforts to challenge current power relations must involve processes of education and communications.

Think about the ways that we are influenced by advertising to buy the 'perfect tomato' (which may be grown with pesticides and transported thousand of miles to our table); or how kids have been seduced into fast food restaurants and childhood obesity by the toys offered along with the fries (Barndt, 2007). This is hegemony at work, reflecting our consent to eat food that may not be good for us. Persuaded by advertising, we are also inadvertently consenting to an unsustainable food system. Except that most people don't make the connection. Nor do they think they have many options, other than the cheap and convenient, given that most of us have been deskilled in growing and cooking our own food.

FoodShare's Recipe for Change program (www.foodshare.net) is tackling that tremendous gap directly by advocating that the provincial Ministry of Education integrate food literacy into the entire primary and secondary school curriculum; this would require that all students learn how to grow and cook food, while also learning the political, economic, ecological, and cultural dynamics of food production and consumption. Headed by a graduate of the Faculty of Environmental Studies (FES), the campaign could be supported by the Bondar report promoting the integration of environmental education across all subjects as well as the Equity policy of the Toronto Board of Education.

This suggests a popular education approach that links the daily experiences of the students, their families and communities with the broader history of food and agriculture, the industrialized food system and climate change, the diverse cultural practices of immigrants, and the community-based initiatives promoting climate justice. Food is a great code for an interdisciplinary education, for experiential learning, and for linking theory with practice — all features of popular education. The FoodShed Project is committed to working with partner organizations to promote nonformal education with communities as well as public school transformation.

Rooted in passion: community arts

Popular education and community arts are counter-hegemonic practices within the cultural sphere. Both practices are about engaging minds, hearts, and bodies in transformative processes, which aim to develop critical social consciousness and move toward more collective actions. While the term 'community arts' is relatively new, the process it refers to — the engagement of people in representing their collective identities, histories, and aspirations in multiple forms of expression — is as old as cave paintings and ritualistic chanting. In Aboriginal contexts, art was/is seen as "an expression of life" practiced by all the people, integrated into ceremony and community (Cajete, 1994: 154).

The separation of 'art' from 'community' perhaps has its roots in both a body/mind and a nature/culture split in Western consciousness emerging from certain streams of the European scientific revolution of the 1700s (Griffin, 1995) and in the commodification of art and knowledge associated with industrial capitalism of the 1800s (Berger, 1972). This has intensified in recent decades with commercialized and individualistic practices of art and media in the context of corporate cultural globalization, often "reducing culture to commerce" (Adams & Goldbard, 2002: 20). In fact, the same forces that have deskilled us in food production have also made people feel that they aren't creative or able to make art that reflects their lives.

We can draw, however, from examples of community-engaged art and media in historical social movements. Community development and community animation in the radical 1960s (*animation socioculturelle* in Quebec) linked the organizing of marginalized communities with the expression of their issues through theatre such as Teatro Campesino (Rose-Avila, 2003), video such as the Canadian National Film Board's Challenge for Change program (Marchessault, 1995), and music such as Black spirituals in the civil rights movement (Sapp, 1995).

Among activists of the new millennium, there has been a resurgence of participatory production of the arts, often in response to the commodified culture of global capitalism and the promotion of passive consumption rather than active citizenship. It is evident in the proliferation of puppets, masks, and performance artists in street protests (Hutcheson, 2006), as well as in the adoption of culture jamming practices (Liacas, 2005), theatre of the oppressed techniques (Boal, 2001), hip hop music, and reclaim the streets movements (Jordan, 2002). It is perhaps most fertile currently in creative activist art blossoming from multiple sites through new social media and web-based activisms (Kidd, 2005).

Cleveland (2002) suggests that community arts can nurture four different kinds of purposes: to educate and inform us about ourselves and the world; to inspire and mobilize individuals and groups; to nurture and heal people and/or communities; and to build and improve community capacity. The social experience of art-making can open up aspects of peoples' beings, their stories, their memories and aspirations, in ways that other methods might miss. When people are given the opportunity to tell their own stories — whether through oral traditions, theatre, visual arts, music, or other media — they bring their bodies, minds, and spirits into a process of communicating and sharing their experiences; they affirm their lives as sources of knowledge, and they stimulate each other in a synergistic process of collective knowledge production.

Community arts are central to the FoodShed Project, as partners are eager to find ways to tell the stories of their innovative practices and to inspire a broader public. We have been experimenting with a new community media process of digital storytelling. In the winter of 2010, York graduate students in a cultural production workshop collaborated with FoodShed partner organizations to critically explore the contradictions around 'local food and food justice.' Adapting the digital storytelling method in various ways, students produced short videos featuring personal narratives by organization members that revealed central principles and challenges of the local food movement (Figure 3). These short videos may be used by the groups themselves, in all-candidates meetings around the municipal elections, in schools with a teacher's guide, and so forth.

Figure 3: York public seminar linking local food groups with digital storytellers (poster by Todd Barsanti/photo by author).

York students in the Community Arts Practice (CAP) program (www.yorku.ca/cap) also collaborated with local food and art groups in organizing a community event as part of the annual Eco Art and Media Festival. "Growing Art, Rooted in Communities" involved The Stop Community Food Centre, the Association of Native and Performing Arts, the Latin American Canadian Art Projects and the Storytelling Centre at the Artscape Wychwood Barns in a series of activities overlapping with the Saturday Farmers' Market. Native singing and drumming filled the barns while market goers contributed to a zine of recipes and stories; family-oriented hands-on art-making workshops produced instruments, placemats, cornhusk dolls, and seed plantings while storytellers and dancers entranced audiences; and students served bread and home-made soups to over 150 community members.

Rooted in praxis: participatory action research

Pure action without reflection is uncritical and nonstrategic activism, while pure reflection without action is mere verbalism.

—Paulo Freire,
Pedagogy of the Oppressed

Implicit in popular education is a process of participatory action research (PAR), engaging learners in an investigation of their own lives in order to more deeply understand the power relations that limit them so they can become more conscious and active agents of change. Participatory research in fact originated within popular education networks, and is understood to be integral to the three-pronged process of research, education, and action associated with Freirean-shaped popular education. Community arts and popular communications are thus tools in this process of people researching their own lives.

According to Carr and Kemmis (1986), PAR is critical social research, different from positivist research (often carried out by a detached scientist) or interpretive research (focusing on subjective meanings). The purpose of liberatory or critical research is the creation of movement for personal and social transformation in order to redress injustices, support peace, and form spaces of democracy. PAR is thus distinguishable from other forms of research by its action component and by being carried out on a group basis (rather than by external researchers independently). It involves *praxis*, or reflecting on what needs to be done, taking action, and reflecting on that action.

PAR has also been a fertile ground for the development of arts-based research methods, a growing field within education, health and social science research. Susan Finley locates 'arts-based research' in the realm of socially transformative approaches: "By its integration of multiple methodologies used in the arts with the postmodern ethics of participative, action-oriented, and politically situated perspectives for human social inquiry, arts-based inquiry has the potential to facilitate critical race, Indigenous, queer, feminist and border theories and research methodologies" (Finley, 2008: 71).

The FoodShed Project has adopted participatory action research as a primary methodology, as it honours the diverse collaborating groups as equal partners in determining the themes, forms, and actions we take. A major goal of the project is to recover the histories of these innovative groups — facilitating collective reflection on their organizational ecology as well as how they engage in both generational and cultural renewal.

Project partners have been experimenting with PAR over the past decade. While working as the Urban Agriculture coordinator for Foodshare in the late 1990s, Lauren Baker facilitated a participatory research process with eight ethnic gardens in Toronto; the Seeds of Our City project (Baker, 2002) involved gardeners in researching the productivity of their diverse community garden practices and exchanging both knowledge and seeds with each other. Baker, FES grad and now executive director of Sustain Ontario, a provincial network for healthy food and farming, recently published a food policy document commissioned by the Metcalf Foundation. Employing a popular education approach, *Menu 2020: Ten Good Food Ideas for Ontario*, advocated for policy that would support local producers, encourage new and alternative farms, compensate farmers as environmental stewards, provide space for urban growers, create integrated school food policies, fund community food centres, create regionally based food clusters, promote local food procurement, integrate good food into health promotion, and educate planners in agricultural land-use options (Hui, 2010: A3). The link between research and action is not always direct, but rather cumulative; these proposals for policy change have come out of myriad gatherings of a growing food network in the city and province.

Dig where you stand: reimagining university education

If we are to dig where we stand, let's end by considering how university education might be rooted in place, politics, passion and praxis. Can we see a place for processes of decolonization, popular education, community arts, and participatory action research in our classrooms and university community?

At York University, we find ourselves standing on a shifting terrain, amidst several troubling trends:

- The deepening corporatization of the neoliberal university, with eroding public funding and increasingly more dependence on private monies and interests;
- The prioritizing of federal and provincial monies for science and technology, business and law, with diminishing support for the humanities, arts and social sciences;
- The measurement of successful teaching reduced to the number of 'bums in seats' and of successful research reduced to the amount of tricouncil research monies secured;
- An increasingly market-driven curriculum which assumes that obtaining a well-paying job is the ultimate mission of universities in the 21st century.

York was founded over 50 years ago with a commitment to social justice. It proudly touts the most diverse student body in

Deborah Barndt

the country, and promotes itself as the interdisciplinary university. A recent White Paper,[4] developed through consultation with the York community, proposes the new moniker of 'the engaged university,' highlighting community-university collaboration and experiential learning. Now here's the Big Chill: inevitably, these goals collide directly with the trends outlined above. Philosopher Martha Nussbaum, in a recent interview in *The Globe and Mail* warns that the "growing obsession with knowledge that you can take to the bank" that eschews the teaching of critical thinking for quick fix technical training is coming back to haunt us. It has in fact led to the Wall Street economic collapse and the BP oil spill in the Gulf of Mexico; it also dominates the thinking around climate change, which too often focuses on technical and market solutions to a much deeper problem. The humanities, arts and social sciences are essential, Nussbaum contends, so students aren't "just accepting what's passed down from some kind of authority, but thinking critically about it, examining yourself and figuring out what you really want to stand for" (Nussbaum in Allemang, 2010: F1). They bring richer meaning to life, nurture an empathetic understanding of differences among people, and train the imagination.

In spite of the chilly climate of the neoliberal university, there are initiatives at York that are very relevant to the FoodShed project goals, and epitomize the mantras of experiential learning, community engagement, and social justice as well as sustainability. In the 1990s, FES students started the Maloca Garden, a community garden at the southwest corner of the campus. While it remains a symbolic gesture toward relocalizing production, the President's Sustainability Council, a Sustainable Purchasing Coalition, and the Institute for Research on Sustainability are all engaged in exploring ways that York can move its food services toward more sustainable and just practices. A new Zero Waste policy is also a step in the direction toward climate justice. Along with promoting a deeper analysis of climate change framed by a climate justice perspective, students can get involved in these and other initiatives, through class assignments, independent studies, and extracurricular activity.

They, too, can dig where they stand...!

Dig Where You Stand!

* * *

Deborah Barndt *is a Professor in the Faculty of Environmental Studies at York University.*

ENDNOTES

[1] The FoodShed Project is coordinated by Harriet Friedmann of the University of Toronto and the author.

[2] http://www.edu.gov.on.ca/curriculumcouncil/shapingSchools.html, accessed, 26 June 2010.

[3] http://www.tdsb.on.ca/_site/viewitem.asp?siteid=15&menuid=682&pageid=546, accessed 26 June 2010.

[4] http://www.yorku.ca/vpaweb/whitepaper/accessed, 26 June 2010.

REFERENCES

Adams, Don and Arlene Goldbard, 2002. *Community, Culture, and Globalization*. New York: The Rockefeller Foundation.

Allemang, John, 2010. "Socrates Would Be So Proud," *The Globe and Mail*, 12 June, F1.

Baker, Lauren, 2002. *Seeds Of Our City: Case Studies From Eight Diverse Community Gardens*. Toronto: Foodshare Education and Research Office.

Barndt, Deborah, 2007. *Tangled Routes: Women, Work and Globalization on the Tomato Trail*. Lanham, MD: Rowman & Littlefield.

Berger, John, 1972. *Ways of Seeing*. London: Penguin Books.

Boal, Augosto, 2002. *Games for Actors and Non-Actors*, Second edition, Adrian Jackson, translator. New York: Routledge.

Cajete, Gregory, 1994. *Look to the mountain: An ecology of Indigenous education*. Kyland, NC: Kivaki Press.

Carr, W. and S. Kemmis, 1986. *Becoming Critical. Education, knowledge and action research*, Lewes, UK: Falmer Press.

Cleveland, William, 2002. "Mapping the Field: Arts-Based Community Development." http://www.communityarts.net/readingroom/archive/intro-develop.php, accessed 26 June 2010.

Finley, Susan, 2008. "Arts-Based Research," p. 71. In G. Knowles and A. Cole, editors, *Handbook of the Arts in Qualitative Research*. Thousand Oaks, CA: SAGE, Publications.

Freire, Paulo, 1982. *Pedagogy of the Oppressed*. New York: Continuum.

Gramsci, Antonio, 1971. *Selections from the Prison Notebooks of Antonio Gramsci*. New York: International Publishers.

Griffin, Susan, 1995. *The Eros of Everyday Life: Essays on Ecology, Gender and Society*. Toronto: Doubleday.

Hui, Ann, 2010. "Setting the table to beat the 'good food gap,'" *The Globe and Mail*, 16 June, A3.

Hutcheson, Maggie, 2006. "Demechanizing Our Politics: Street Performance and Making Change," pp. 79-87. In Deborah Barndt, editor, *Wild Fire: Art as Activism*. Toronto: Sumach Press.

Jordan, John, 2002. "The Art of Necessity: The Subversive Imagination of Anti-road Protest and Reclaim the Streets," pp. 347-57. In Stephen Duncombe, editor, *Cultural Resistance Reader*. London: Verso.

Kidd, Dorothy, 2005. "Linking Back, Looking Forward," pp. 151-61. In Andrea Langlois & Frederic Dubois, editors, *Autonomous Media: Activating Resistance and Dissent*. Montreal: Cumulus Press.

Liacas, T., 2005. "101 Tricks to Play with the Mainstream: Culture Jamming as Subversive Recreation," pp. 61-73. In Andrea Langlois & Frederic Dubois, editors, *Autonomous Media: Activating Resistance and Dissent*. Montreal: Cumulus Press.

Lindqvist , Sven, 1979. "Dig Where You Stand," *Oral History*, 7, 2, Autumn, 24-30.

Loomba, Ania, 1998. *Colonialism / Postcolonialism*. New York: Routledge.

Marchessault, J., 1995. Reflections on the dispossessed: Video and the 'Challenge for Change' Experiment, *Screen*, 36, 2 (Summer), 131-146.

Roberts, Wayne, 2008. *A No-Nonsense Guide to World Food*. Toronto: Between the Lines.

Rose-Avila, Magdaleno, 2003. "Homegrown Revolution," *ColorLines: Race Culture Action*, Fall, 10-12.

Said, Edward, 2003, p. 59. In Robert Young, *Postcolonialism: A Very Short Introduction*. Oxford, UK: Oxford University Press.

Sapp, Jane, 1995. "To Move and To Change," *ARTS: The Arts in Religious and Theological Studies*, 7, 2, 30-33.

Spivak, Gayatri, 1988. "Can the Subaltern Speak?" pp. 24-28. In Cary Nelson and Lawrence Grossberg, editors, *Marxism and the Interpretation of Culture*. Champaign, IL: University of Illinois Press.

Tuhiwai Smith, Linda, 1999. *Decolonizing Methodologies: Research and Indigenous Peoples*. New York: Zed Books.